CITES and Slipper Orchids

An introduction to slipper orchids covered by the Convention on International Trade in Endangered Species

Written by

H. Noel McGough,

David L. Roberts, Chris Brodie and Jenny Kowalczyk

Royal Botanic Gardens, Kew

United Kingdom

The Board of Trustees, Royal Botanic Gardens, Kew

2006

First published in 2006 by
Royal Botanic Gardens, Kew
Richmond, Surrey, TW9 3AB, UK
www.kew.org

ISBN 1-84246-128-1

For information or to purchase Kew titles please visit
www.kewbooks.com or email publishing@kew.org

CONTENTS

INTRODUCTION

The aim of '*CITES and Slipper Orchids*' is to provide an introduction to the slipper orchids on CITES. This includes identification, trade, and thc implementation of CITES for slipper orchids.

The guide is primarily intended as a training tool for those working with the Convention, namely CITES Management Authorities, Scientific Authorities and enforcement agencies. However, it is also likely to be of interest to a much wider audience, especially those interested in how CITES works for this commercially important group of plants.

'*CITES and Slipper Orchids*' has been designed so that it can be easily adapted to match the needs of the presenter. We encourage the user to 'tailor' the presentation to suit their audience. In addition to the speaker's notes we have provided a bibliography and list of resources. We hope you find the pack not only a useful tool for developing your presentations, but also a convenient reference book. Please use this training tool and feed back your comments to us so that we can revise future editions to suit your needs.

Noel McGough,

Head of the Conventions and Policy Section,

UK CITES Scientific Authority for Plants,

Royal Botanic Gardens, Kew

ACKNOWLEDGEMENTS

The authors wish to thank the following for their expertise and guidance in the preparation of this pack: Wendy Byrnes, Phillip Cribb, Margarita Clemente Muñoz, Kate Davis, Deborah Rhoads Lyon, Rosemary Simpson, Matthew Smith, Sabina Michnowicz, and Ger van Vliet.

The pack was funded by the United Kingdom's CITES Management Authority, the Department for Environment, Food and Rural Affairs (Defra).

HOW TO USE THIS PRESENTATION PACK

This pack consists of slides and speaker's notes for a presentation on the slipper orchids listed in the CITES Appendices. The presentation is divided into three separate topic areas that can be used and adapted according to the background, interests and needs of your audience (Introduction to Slipper Orchids, Slipper Orchids on CITES, Implementing CITES for Slipper Orchids).

A fourth section of additional slides and speaker's notes provides detail on some extra topics that you can add to your presentation, as you think appropriate. The slides have been drafted in general terms with the hope that they will remain current, and therefore of use, for the foreseeable future.

Suggested speaker's notes accompany each slide. These notes are more specific than the slides and reflect information current as of May 2005. Of course, all speakers are encouraged to express their personal style and to use notes as closely or as loosely as they feel comfortable!

We hope that this pack will provide a useful starting point from which you can tailor the slides, and accompanying speaker's notes, to reflect the specific needs of your audience, the length of the presentation and your own personal style. For example, you could illustrate some slides with examples from your own region or institution, or supplement the slides with extra images, such as cartoons, photographs, or newspaper cuttings. Such measures will undoubtedly increase the impact of an individual presentation. In addition, the slides can be printed onto transparency sheets for use with an overhead projector. Alternatively, they can be photocopied from this book, or printed as handouts from the Microsoft® PowerPoint® file on the CD-ROM, and given out to an audience for information.

CD-ROM

The CD-ROM contains the following files:

'CITESSlipperOrchids.ppt', a Microsoft PowerPoint® presentation containing the slides and speaker's notes. You will need Microsoft PowerPoint 97® (or a more recent version) installed on your computer to view and customise this file.

'CITESSlipperOrchids.pdf', an Adobe Acrobat® presentation. You cannot modify this presentation but it can be viewed in "full screen" mode using Adobe Reader®. You will need Adobe Acrobat Reader® installed on your computer to view this file (can be downloaded from www.adobe.com).

'CITESSlipperOrchidsBW.pdf', an Adobe Acrobat® presentation in black and white (can be downloaded from www.adobe.com).

'CITESSlipperOrchidsPack.pdf', a full copy of the text for the pack including the introduction, references and speaker's notes. This allows you to view the complete electronic document as well as print off part or all of the pack. You will need Adobe Reader® installed on your computer to view this file (can be downloaded from www.adobe.com).

REFERENCES AND RESOURCES

References to the Convention

CITES (2003 and updates). *CITES Handbook.* Secretariat of the Convention on International Trade in Endangered Species of Wild Fauna and Flora, Geneva, Switzerland. This handbook includes the text of the Convention and its Appendices, a copy of a standard permit and the text of the Resolutions and Decisions of the Conference of the Parties.

Wijnstekers, W. (2003 and updates). *The Evolution of CITES, 6th edition.* Secretariat of the Convention on International Trade in Endangered Species of Wild Fauna and Flora, Geneva, Switzerland. The most comprehensive and authoritative reference available to the Convention, written by the CITES Secretary General. Updated on a regular basis.

Rosser, A. and Haywood, M. (Compilers), (2002). *Guidance for CITES Scientific Authorities. Checklist to assist in making non-detriment findings of Appendix II exports.* Occasional Paper of the IUCN Species Survival Commission No. 27. IUCN - The World Conservation Union, Gland, Switzerland and Cambridge, United Kingdom. The first attempt to define guidelines to be used by Scientific Authorities when they make the non-detriment statement required before issuance of a CITES export permit.

The CITES website (www.cites.org) contains a wide variety of information on the Convention: species listed in the Appendices, key addresses and contacts, reports of meetings and working groups, new publications and websites and a diary of events.

Reviews of the Convention

Hutton, J. and Dickson, B. (2000). *Endangered Species, Threatened Convention. The Past, Present and Future of CITES.* Earthscan, London, United Kingdom. A critical assessment of CITES from the sustainable use perspective.

Oldfield, S. (Editor), (2003). *The Trade in Wildlife: Regulation for Conservation.* Earthscan, London, United Kingdom. Critically reviews the international trade in wildlife.

Reeve, R. (2002*). Policing International Trade in Endangered Species. The CITES Treaty and Compliance.* Royal Institute of International Affairs. Earthscan, London, United Kingdom. A detailed study of the CITES compliance system.

CITES Standard References for Plants - Checklists

Carter, S. and Eggli, U. (2003). *The CITES Checklist of Succulent Euphorbia Taxa (Euphorbiaceae).* Second edition. German Federal Agency for Nature Conservation, Bonn, Germany. Reference to the names of succulent *Euphorbia.*

Hunt, D. (1999). *CITES Cactaceae Checklist.* Second edition. Royal Botanic Gardens, Kew, United Kingdom. Reference to the names of *Cactaceae*, the cactus family.

Mabberley, D.J. (1997). *The Plant-Book.* Second edition. Cambridge University Press, Cambridge, United Kingdom. The reference for the generic names of all

CITES plants, unless they are superseded by standard checklists adopted by the Parties as referenced in this list.

Newton, L.E. and Rowley, G.D. (Eggli, U. Editor), (2001). *CITES Aloe and Pachypodium Checklist.* Royal Botanic Gardens, Kew, United Kingdom. Reference for the names of *Aloe* and *Pachypodium.*

Roberts, J.A., Beale, C.R., Benseler, J.C., McGough, H.N. and Zappi, D.C. (1995). *CITES Orchid Checklist. Volume 1.* Royal Botanic Gardens, Kew, United Kingdom. Reference to the names of *Cattleya, Cypripedium, Laelia, Paphiopedilum, Phalaenopsis, Phragmipedium, Pleione* and *Sophronitis* including accounts of *Constantia, Paraphalaenopsis* and *Sophronitella.*

Roberts, J.A., Allman, L.R., Beale, C.R., Butter, R.W., Crook, K.B. and McGough, H.N. (1997). *CITES Orchid Checklist. Volume 2.* Royal Botanic Gardens, Kew, United Kingdom. Reference to the names of *Cymbidium, Dendrobium, Disa, Dracula* and *Encyclia.*

Roberts, J.A., Anuku, S., Burdon, J. , Mathew, P., McGough, H.N. and Newman, A.D. (2001). *CITES Orchid Checklist. Volume 3.* Royal Botanic Gardens, Kew, United Kingdom. Reference to the names of *Aerangis, Angraecum, Ascocentrum, Bletilla, Brassavola, Calanthe, Catasetum, Miltonia, Miltonioides, Miltoniopsis, Renanthera, Renantherella, Rhynchostylis, Rossioglossum, Vanda* and *Vandopsis.*

Willis, J.C., revised by Airy Shaw, H.K. (1973). *A Dictionary of Flowering Plants and Ferns.* 8th edition. Cambridge University Press. Cambridge, United Kingdom. For generic synonyms not mentioned in The Plant-Book, unless they are superseded by standard checklists adopted by the CITES Parties as referenced in this list.

UNEP-WCMC (2005). *Checklist of CITES Species.* UNEP-WCMC, Cambridge, United Kingdom. The COP has adopted this Checklist and its updates as an official digest of scientific names contained in the standard references.

CITES Checklists are updated on a regular basis by the CITES Nomenclature Committee. See the CITES website for more information: www.cites.org.

Further References

The following are general references which we hope you will find useful. Be aware that the taxonomy in these works may differ from that prescribed in the adopted CITES references given above. Please inform us of works that you find useful and we will include them in future editions of the guide.

Averyanov, L., Cribb, P., Loc P.K. & Hiep. N.H. (2003). *Slipper Orchids of Vietnam.* Royal Botanic Gardens, Kew, United Kingdom. Comprehensive review with full description of all species of *Paphiopedilum* native to Vietnam, line drawings and extensive use of colour photographs including some habitat pictures.

Bechtel, H. Cribb, P. and Launert, E. (1992). *The Manual of Cultivated Orchid Species.* Third Edition. Blandford Press, London, United Kingdom. Badly in need of updating but still an excellent reference. Has detailed analysis of over 400 genera and 1,200 species with over 860 colour photographs and many fine line drawings.

References and Resources

Braem, G. J., Baker, C.O and Baker, M.L. (1998). *The Genus Paphiopedilum: Natural History and Cultivation, Volume 1.* Botanical Publishers Inc., Kissimmee, Florida, USA.

Braem, G. J., Baker, C.O and Baker, M.L. (1999). *The Genus Paphiopedilum: Natural History and Cultivation, Volume 2.* Botanical Publishers, Inc., Kissimmee, Florida, USA.

Braem, G. J. and Chiron, G.R. (2003). *Paphiopedilum.* Tropicalia, Saint-Genis Laval, France.

Cash, C. (1991). *The Slipper Orchids.* Christopher Helm, London, United Kingdom.

Cavestro, W. (2001). *Le genre Paphiopedilum: taxonomie, répartition, habitat, hybridation et culture.* Rhône-Alpes Orchidées, Lyon, France.

Chen, V.Y. and Song, M. (2000). *Guide to CITES Plants in Trade* (Chinese edition). TRAFFIC East Asia.

CITES (1993-). *CITES Identification Manual, Volume 1 Flora.* Secretariat of the Convention on International Trade in Endangered Species of Wild Fauna and Flora. Geneva, Switzerland. This is the official CITES identification manual. Parties are required to produce sheets for the manual if they successfully propose a species for listing. New identification sheets are added to this ring-bound manual on an ongoing basis. Essential for anyone working on CITES and plants.

Cribb, P. (1997). *Slipper Orchids of Borneo.* Natural History Publications (Borneo), Kota Kinabalu, Malaysia.

Cribb, P. (1997). *The Genus Cypripedium - A Botanical Magazine Monograph.* Published in association with the Royal Botanic Gardens, Kew. Timber Press, Portland, USA. Comprehensive monograph with full description of all species, line drawings, colour photographs and colour paintings.

Cribb, P. (1998). *The Genus Paphiopedilum (Second Edition) - A Botanical Magazine Monograph.* Published in association with the Royal Botanic Gardens, Kew. Natural History Publications (Borneo), Kota Kinabalu, Malaysia. Comprehensive monograph with full description of all species, line drawings, colour photographs and colour paintings.

European Commission (2002). *Five years of new wildlife trade regulations.* Office for Official Publications of the European Communities, Luxembourg. Booklet on the EU wildlife trade regulations.

Hennessy, E. F. and Hedge, T.A. (1989). *The Slipper Orchids.* Acorn Books, Randburg, RSA.

Gruss, O. (2003). *A Checklist of the Genus Phragmipedium.* Orchid Digest 67[4]: 213-255.

Hilton-Taylor, C. (Compiler), (2000-). *IUCN Red List of Threatened Species.* IUCN-The World Conservation Union, Gland, Switzerland and Cambridge, United Kingdom. The official IUCN list of threatened plants and animals, published as a booklet with CD-ROM. The list is constantly being updated and improved. For the latest version check the Red List website on www.redlist.org.

IUCN/SSC Orchid Specialist Group. (1996). *Orchids – Status Survey and Conservation Action Plan.* IUCN, Gland, Switzerland and Cambridge, United Kingdom.

Jenkins, M. and Oldfield, S. (1992). *Wild Plants in Trade.* TRAFFIC International, Cambridge, United Kingdom. A summary of the last full survey of European CITES plant trade.

Koopowitz, H. (2000). *A revised checklist of the Genus Paphiopedilum.* Orchid Digest 64[4]: 155-179.

Lange, D. and Schippmann, U. (1999). *Checklist of medicinal and aromatic plants and their trade names covered by CITES and EU Regulation 2307/98 Version 3.0.* German Federal Agency for Nature Conservation, Bonn, Germany.

Marshall, N.T. (1993). *The Gardener's Guide to Plant Conservation.* TRAFFIC North America. Unfortunately now dated, although there are rumours of a new edition in the pipeline. This was a very useful guide to plants in trade for horticulture and their sources.

Mathew, B. (1994). *CITES Guide to Plants in Trade*. Department of the Environment, London, United Kingdom. Now dated, but contains colour photographs and descriptions of the major CITES plant groups controlled and traded in the early 1990s.

McCook, L. (1998). *An annotated checklist of the genus Phragmipedium.* Orchid Digest Corp., CA, USA. Special publication of the Orchid Digest.

McGough, H.N., Groves M., Mustard M and Brodie, C. (2004), *CITES and Plants - A User's Guide.* Royal Botanic Gardens, Kew, United Kingdom.

Pridgeon, A. (2003). *The Illustrated Encyclopedia of Orchids.* David and Charles, Devon, United Kingdom. Over 1100 species illustrated. Outlines the major taxa in trade and of interest to collectors. Full colour photographs. The best general guide to orchids available in print.

Rittershausen, W. & B. (1999). *Orchids – a practical guide to the world's most fascinating plants.* The Royal Horticultural Society, reprinted 2004. Quadrille Publishing Ltd, London, United Kingdom.

Sandison, M.S., Clemente Muñoz, M., de Koning J. and Sajeva, M. (1999). *CITES and Plants - A User's Guide.* Royal Botanic Gardens, Kew, United Kingdom. First 'slide pack' of 40 slides and text produced in English, French and Spanish.

Sandison, M.S., Clemente Muñoz, M., de Koning J. and Sajeva, M. (2000). *CITES and Plants - A User's Guide*. (Chinese Edition). Royal Botanic Gardens, Kew, United Kingdom. Edited by Vincent Y. Chen and Michael Song and produced by TRAFFIC East Asia. The User's Guide in Chinese.

Schippmann, U. (2001). *Medicinal Plants Significant Trade Study CITES Project S-109. Plants Committee Document PC9 9.1.3(rev.). BfN - Skripten 39.* German Federal Agency for Nature Conservation, Bonn, Germany. An excellent overview of the trade in CITES-listed medicinal plants.

CD-ROM

CITES (2002-). *CITES training presentations.* CITES Secretariat, Geneva, Switzerland. A range of training presentations produced by the Capacity Building Unit of the CITES Secretariat. These are essential tools for anyone carrying out CITES training.

CITES (2003-). *CD-ROM version of the CITES website* (*www.cites.org*). Full version of the CITES website on CD-ROM. Available from the CITES Secretariat.

Web Sites

There are a large number of sites of some interest to CITES workers. Many national CITES authorities have their own dedicated websites. The following are key sites that will lead you to as many other sites as you have time to spend on the Web.

CITES Home Page: Official site of the CITES Secretariat. Includes lists of Parties, Resolutions and other documents. www.cites.org.

European Commission: Information on the Wildlife Trade Regulations that implement CITES within the European Union. www.eu-wildlifetrade.org.

UK CITES Website: Website maintained by the UK CITES authorities that aims to provide information and updates on CITES-related matters as they pertain to the United Kingdom and its Overseas Territories. www.ukcites.gov.uk.

IUCN - The World Conservation Union: The world's largest professional conservation organisation. IUCN brings together governments, non-governmental organisations, institutions and individuals to help nations make the best use of their natural resources in a sustainable manner. www.iucn.org.

IUCN Species Survival Commission: SSC is the IUCN's foremost source of scientific and technical information for the conservation of endangered and vulnerable species of flora and fauna. Specific tasks are carried out on behalf of IUCN, such as the monitoring of vulnerable species and their populations, the implementation and review of conservation action plans and the provision of guidelines, advice and policy recommendations to governments, agencies and organisations regarding conservation and management of species and their populations. www.iucn.org/themes/ssc/.

UNEP - World Conservation Monitoring Centre: The UNEP-WCMC provides information services on the conservation and sustainable use of the world's living resources, and assists others in the development of information systems. Activities include supporting the CITES Secretariat. Information on international wildlife trade and trade statistics may be requested from the Species Programme of UNEP - WCMC. Now an office of the UN based in Cambridge, UK, the Centre's work is an integral part of the United Nations Environment Programme (UNEP), headquartered in Nairobi, Kenya. www.unep-wcmc.org/index.html.

TRAFFIC International: TRAFFIC is a programme of WWF and the IUCN established to monitor the trade in wild plants and animals. The TRAFFIC Network is the world's largest monitoring programme with offices covering most parts of the world. The network works closely with the CITES Secretariat. www.traffic.org.

Earth Negotiations Bulletin: Tracks the major environmental negotiations as they happen. Also provides an extensive archive material and lots of photographs of the meetings. www.iisd.ca.

Plant Name Checking

The following websites are useful for checking plant names that are not found in the standard CITES checklists. Sometimes these names may be of newly described species. If this 'new name' has been used in an application for a CITES permit stating the plant is propagated, the plant should be checked to confirm its identity and to ensure it meets the CITES definition of artificial propagation.

IPNI - The International Plant Names Index: A database of the names and associated basic bibliographical details of all seed plants. www.ipni.org/index.html.

TROPICOS: A nomenclatural database produced and maintained by the Missouri Botanical Garden, USA. mobot.mobot.org/W3T/Search/vast.html.

EPIC – Electronic Plant Information Centre: Brings together all the digitised information on plants held by the Royal Botanic Gardens, Kew. www.rbgkew.org.uk/epic/.

World Checklist of Monocotyledons – Comprises an inventory of taxonomically validated Monocotyledon plant names and associated bibliographic details, together with their global distribution. This database includes a full list of all orchid names. www.kew.org/monocotChecklist/.

Phragweb - A comprehensive source of information on *Phragmipedium* species, including descriptions, line drawings and full colour photographs. www.Phragweb.info.

RHS – The website of the Royal Horticultural Society, useful for checking new hybrid names. www.rhs.org.uk/publications/pubs_journals_orchid_hybrid.asp.

SLIDE INDEX

Introductory Slides

Introduction to Slipper Orchids

Slipper Orchids on CITES

Implementing CITES for Slipper Orchids

Additional Slides

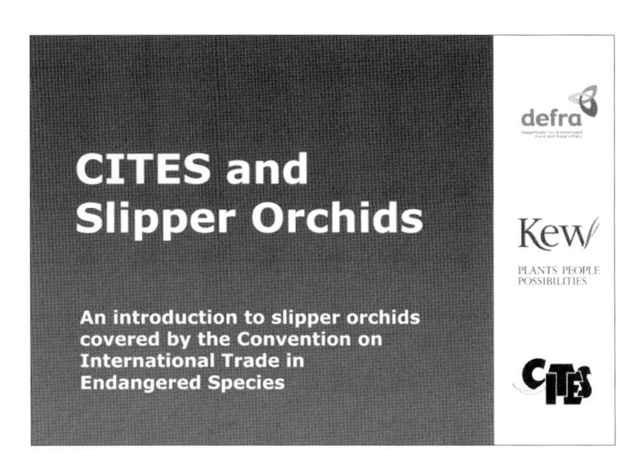

Slide 1: CITES and Slipper Orchids

The aim of this presentation is to introduce you to the different types of slipper orchids covered by the Convention on International Trade in Endangered Species of wild fauna and flora - CITES - and to address some of the key issues concerning the implementation of the Convention for this important plant group.

Slide 2

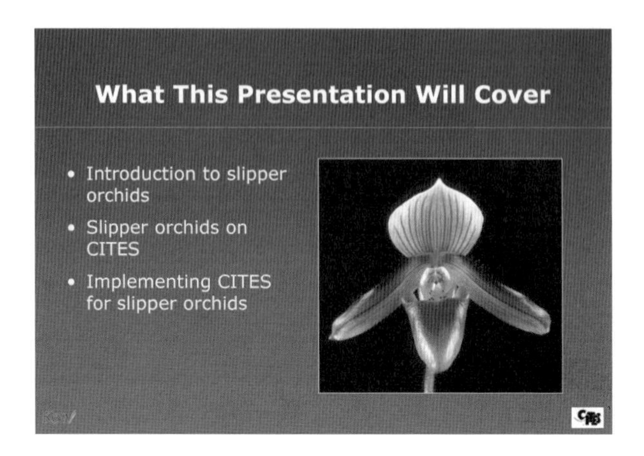

Slide 2: What This Presentation Will Cover

This presentation will cover the following topics:

- an introduction to slipper orchids;
- slipper orchids covered by CITES;
- implementing CITES for slipper orchids.

[Note to speaker: The slide shows Paphiopedilum callosum.]

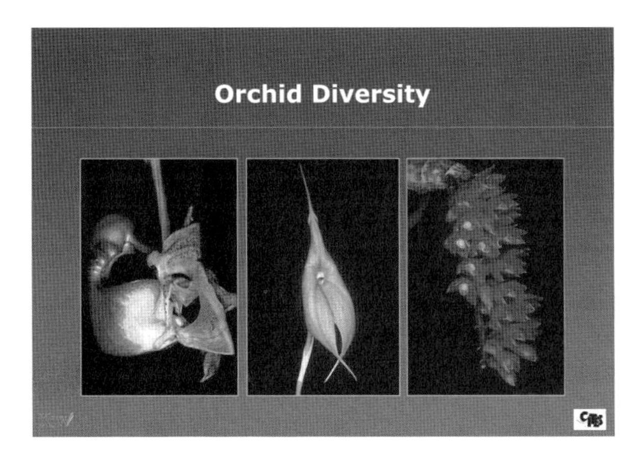

Slide 3: Orchid Diversity

Most people have at least a vague idea of what an orchid is. The name orchid often sparks ideas of fascinating flowers, exotic environments and intriguing gifts! However, the plants that most people see represent just a tiny fraction of what is possibly the largest plant group in the world; with over 25,000 species known and an estimated 5,000 species waiting to be discovered. These species are found all over the world, with their major concentration (some 70%) in the tropical rainforests. However, orchids can even be found in very arid areas and in sub-Arctic areas such as Alaska. Subterranean species have even been found in Australia! Their size can range from that of a pencil head to a vigorous two tonne shrub. Also, contrary to popular belief, not all orchids are rare. In the right habitat, some species can occur in very large numbers. Due to their huge variation in size, colour and their exotic appeal, orchids have fascinated plant collectors and growers since the Victorian era.

[Note to speaker: The characteristics that distinguish orchids from other plants are mostly confined to the flowers. They are:

1. *The male and female flower parts are fused, or at least partly fused, to form a structure called the column;*

2. *One of the flower petals is highly modified, often into a landing platform or guide for pollinating insects - called the labellum or lip;*

3. *The pollen is usually bound together forming large masses called pollinia, which are in pairs, 4s, 6s or 8s;*

4. *Orchids produce millions of very small seeds that have no food source for germination and therefore require a fungal associate to aid germination.*

This slide shows: Coryanthes macrantha (left), Masdevallia veitchiana (centre) and Dendrobium secundum (right).]

Slide 4

Slide 4: Why Protect Orchids?

Orchids are highly threatened by habitat destruction and, to a lesser degree, over-collection. Although habitat destruction affects all species, over-collecting is a particularly serious threat to those species that are important in trade and can lead to the extinction of a species in the wild within a few years of discovery.

Orchids have evolved intricate pollination strategies that can be easily disturbed through habitat destruction or over-collecting - making them good environmental indicators. Therefore, if we can save orchids and maintain good populations, it is likely that we will be saving many other species in the process. Due to their public appeal, orchids also make good flagship species and have been used to promote the conservation of critical wildlife habitats.

[Note to speaker: This slide shows Phragmipedium besseae.]

Introduction to
Slipper Orchids

Slide 6

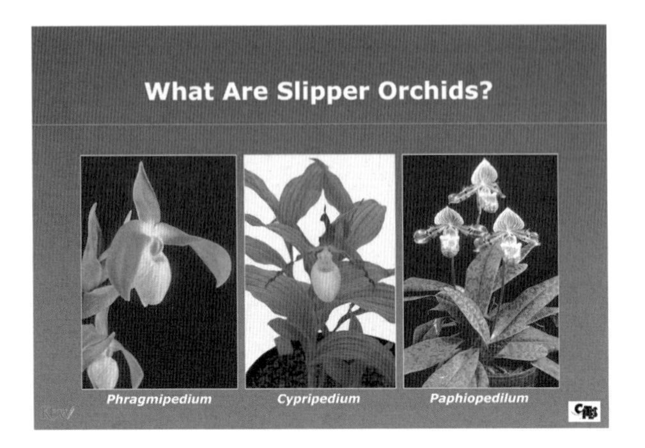

Slide 6: What Are Slipper Orchids?

Slipper orchids are easily distinguished from other orchids by their slipper- or shoe-like flowers. These slipper-like flowers have inspired a range of vernacular names which include Lady's Slipper in Europe, Moccasin Flower in North America and Zapatilla in Latin America. The most important genera of Slipper orchids are *Cypripedium*, *Paphiopedilum* and *Phragmipedium*. All of these genera are in demand for international trade.

The genus *Cypripedium* comprises some 50 species distributed across the northern temperate regions. The genus *Paphiopedilum* consists of some 80 species confined to Southeast Asia, and the genus *Phragmipedium* comprises around 20 species confined to central and South America. Their attractive flowers and relatively small number of species have made these plants highly attractive to orchid collectors and growers.

[Note to speaker: The slide shows Phragmipedium besseae var. dalessandroi (left), Cypripedium parviflorum (centre) and Paphiopedilum venustum (right).]

Slide 7: Slipper Orchids on CITES

The genera *Paphiopedilum* and *Phragmipedium* are listed on Appendix I of CITES. An Appendix I listing effectively prohibits the trade in wild-collected plants, but allows trade in artificially propagated plants, subject to permit. These two genera are listed on CITES Appendix I as *Paphiopedilum* spp. and *Phragmipedium* spp. This 'generic listing' means that newly described species of these genera are automatically included in CITES Appendix I. This ensures the immediate regulation of new species which are vulnerable to non-sustainable trade following their discovery.

The genus *Cypripedium* is listed on Appendix II of CITES. This means that wild and artificially propagated specimens can be traded subject to obtaining the correct permits. However, the majority of range countries (range States in CITES terms) have banned the export of wild-taken *Cypripedium*. In addition, the 25 member countries of the European Union, in their wildlife trade regulations that implement CITES, treats *Cypripedium calceolus* as though it is on Appendix I of the Convention. As a result of these measures it is unusual to find legal wild-collected *Cypripedium* plants in international trade. If legal wild plants are in trade they are usually plants of the commoner North American species that have come from controlled collection or salvage operations.

[Note to speaker: To check the most up-to-date list of the CITES Appendices consult the CITES website www.cites.org. To get more details of the stricter controls implemented by the European Union, log on to the website www.eu-wildlifetrade.org. This slide shows Phragmipedium longifolium (left), Cypripedium himalaicum (centre) and Paphiopedilum henryanum (right).]

Slide 8

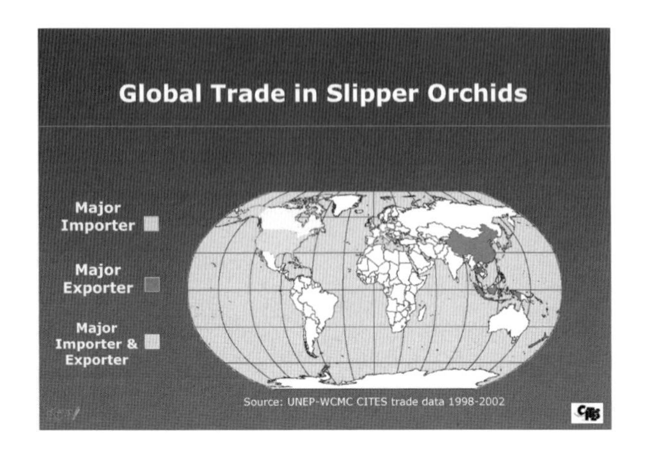

Slide 8: Global Trade in Slipper Orchids

Slipper orchids, namely the genus *Paphiopedilum*, are one of the top 5 most horticulturally important orchid genera. Slipper orchids are traded in large numbers, mainly as living plants of species and man-made hybrids.

The CITES trade data shows that over 660,000 slipper orchids were traded internationally between 1998 and 2002. Almost all of this trade was in artificially propagated material. The recorded trade in wild plants related to specimens of *Cypripedium*.

Between 1998 and 2002, the largest recorded exporters of artificially propagated slipper orchids were Taiwan (Province of China), Indonesia and China. These exporters accounted for over half of all international exports (54%). Other major exporters in the same period were the Republic of Korea, the Netherlands, Thailand, the United States, New Zealand, Japan and Belgium (each exporting more than 10,000 plants).

The largest importer of artificially propagated slipper orchids between 1998 and 2002 was Japan with over half of all imports (56%). Other major importers in the same period were Canada, the Republic of Korea, the United States, Germany, Italy and Switzerland (each importing more than 10,000 plants).

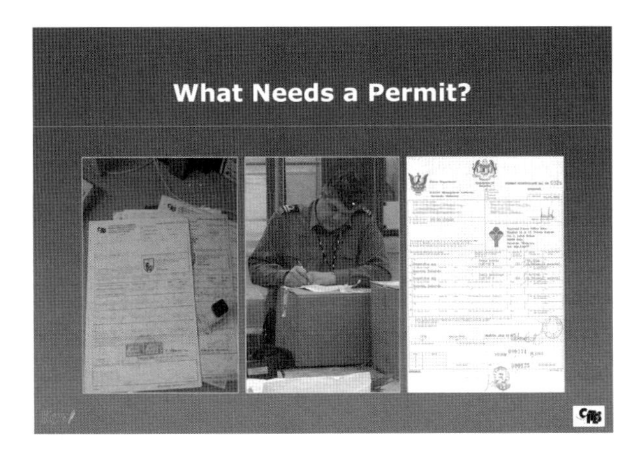

Slide 9: What Needs a Permit?

The simple answer is: everything unless it is exempt!

CITES controls apply to plants, 'alive or dead' and to 'any readily recognisable parts and derivatives'.

This means that it is not just the whole plants that are subject to CITES controls, but also parts of the plants including seeds, cuttings and leaves. Products made from plants may also be subject to CITES controls. If the name of a CITES-listed plant or animal is written on packaging, then the product is considered to contain it and is therefore subject to CITES controls.

For plants listed on CITES Appendix I, the whole plant and all parts and derivatives are controlled - alive or dead. There is only one exemption. Seedling or tissue cultures 'obtained *in vitro,* in solid or liquid media, transported in sterile containers' are exempt. The material need not be in traditional flasks or bottles to be eligible for the exemption – merely in sterile containers. However, it must be of legal origin to fulfil this exemption. We will look in detail at the controls relating to tissue cultures in the next slide.

For plants listed on CITES Appendix II, the plant is controlled 'alive or dead' and so is any readily recognisable part or derivative specified in the appendices. In the case of *Cypripedium* the only parts and derivatives exempted from CITES control are: a) seeds and pollen (including pollinia); b) seedling or tissue cultures obtained *in vitro,* in solid or liquid media, transported in sterile containers; and c) cut flowers of artificially propagated plants.

[Note to speaker: The slide shows CITES permits and a customs officer checking documents.]

Slide 10

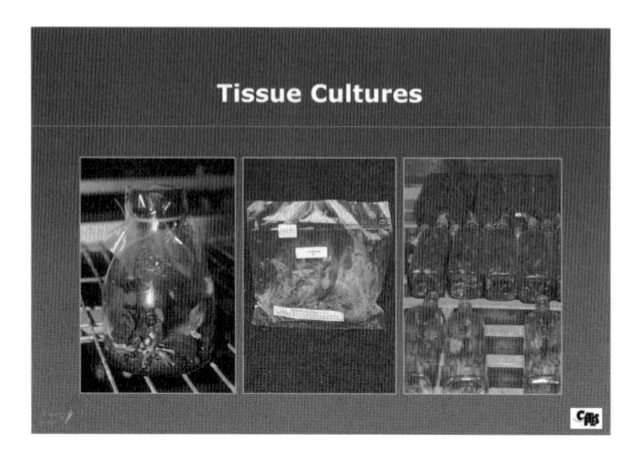

Slide 10: Tissue Cultures

Following the listing of *Paphiopedilum* and *Phragmipedium* on Appendix I, the CITES Conference of the Parties (CoP) approved the exemption of tissue cultures of these genera from CITES control. This was a unique exemption – the removal of a 'readily recognisable part and derivative' of an Appendix I specimen from control. A number of Parties opposed the exemption; however it was approved by a majority vote. The reason for the exemption was to encourage propagation of these highly sought-after plants in a bid to remove pressure from the wild populations.

At the time it was thought that such propagation could never be detrimental to the wild populations. However, in recent years this has proven not to be the case. In particular the illegal and unsustainable collection of species endemic to Viet Nam has caused particular concern. A number of countries have expressed concern that the flasked seedling and tissue culture exemption has been used by unscrupulous traders to 'legalise' trade in material whose parental stock has been taken illegally from the wild.

The parental stock used to produce tissue cultures should be legally obtained, subject to the laws relating to that species in its country of origin. If this 'mother stock' is illegal, then the tissue culture derived from it does not qualify for the exemption from CITES controls. Such tissue culture material may then be subject to confiscation by CITES enforcement authorities.

[Note to speaker: The slide shows sterile seedling cultures in flasks, bags and bottles.]

Slipper Orchids on CITES

Slide 12

Slide 12: *Cypripedium*

The genus *Cypripedium* consists of some 50 species found in the northern temperate region of Asia, Europe and North America, reaching as far south as Honduras, Guatemala and southern China. They grow in a wide range of habitats from coniferous or mixed deciduous woodlands, to marshes and grasslands. They are terrestrial, with leaves that, in most species, grow fresh from the base each year. The flowers are slipper-like and range in colour from green through white and yellow to red and deep purple. *Cypripedium* species appeal to gardeners in temperate climes as all have some degree of cold hardiness and may be grown outside for at least part of the season.

[Note to speaker: The slide shows Cypripedium parviflorum (left), Cypripedium guttatum (top centre), Cypripedium lichiangense (bottom centre) and Cypripedium smithii (right).]

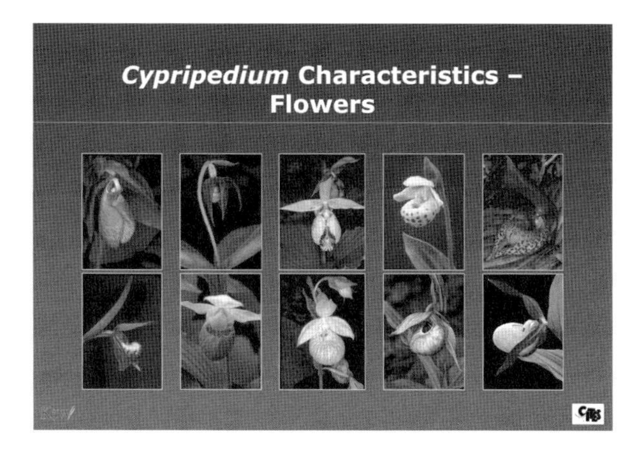

Slide 13: *Cypripedium* Characteristics – Flowers

All *Cypripedium* flowers are slipper-shaped. The flowers can be single to many, with flowers ranging in colour from green to purple-splashed white to golden yellow to purple brown.

The flowers do not fall, but persist on the fruit.

[Note to speaker: The slides shows Cypripedium acaule (top left), Cypripedium palangshanense (second from left, top), Cypripedium japonicum (top centre), Cypripedium wardii (second from right, top), Cypripedium lichiangense (top right), Cypripedium arietinum (bottom left), Cypripedium reginae (second from left, bottom), Cypripedium irapeanum (bottom centre), Cypripedium x froschii (second from right, bottom) and Cypripedium calceolus (bottom right).

Slide 14

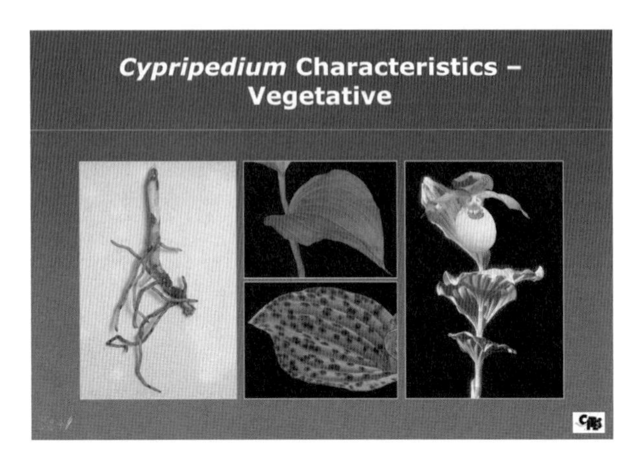

Slide 14: *Cypripedium* Characteristics – Vegetative

- All *Cypripedium* have a characteristic underground root-like modified stem called a rhizome.

- In most species this rhizome is short and seldom branching, producing a rosary bead-like chain of annual growths.

- The leaves die back annually, leaving the rhizome to survive the annual dormant period. In the springtime new growth arises from buds on the rhizome.

- The leaves are usually oval.

- The leaves are pleated on their long axis and have prominent veins.

- The leaves are often hairy, especially on the veins and margins.

[Note to speaker: The slide shows a rhizome with bud and roots (left), leaves showing variation in colour (top and bottom centre), stem leaves and flower of Cypripedium fasciolatum (right).]

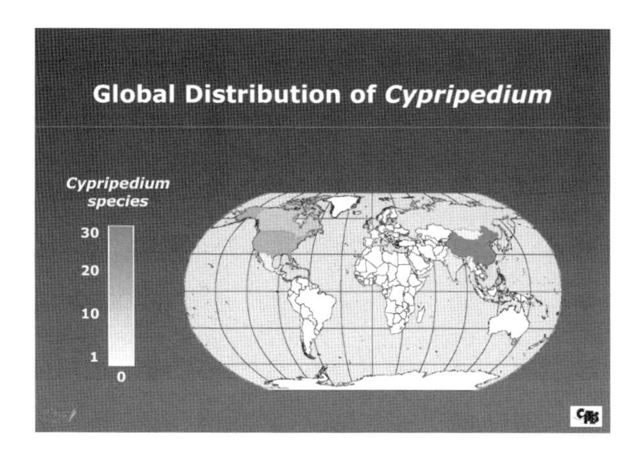

Slide 15: Global Distribution of *Cypripedium*

Although the genus *Cypripedium* is found throughout the northern temperate region, China contains by far the largest number of species. The Chinese species, due to lack of access in the past, are extremely sought-after. Most of the species that are now common in cultivation are those from North America, the European species *Cypripedium calceolus* and the Japanese species *Cypripedium formosanum.*

Slide 16

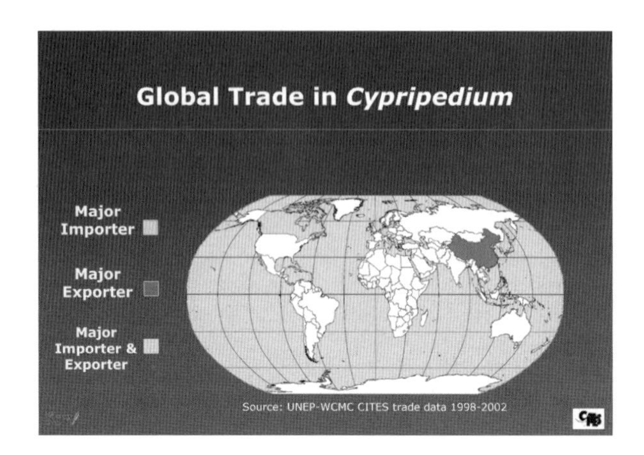

Slide 16: Global Trade in *Cypripedium*

The vast majority of recorded CITES trade in *Cypripedium* taxa for the years 1998-2002 was in live plants. By far the majority of this trade was reported to be in artificially propagated plants (98%).

Taiwan (Province of China), China and the Republic of Korea were the major reported sources of propagated material. These three exporters supplied 93 per cent of reported propagated material between 1998 and 2002.

The main importer of *Cypripedium* taxa between 1998 and 2002 was Japan, accounting for 78 per cent of all CITES recorded imports. Japan, the Republic of Korea, Canada and Germany accounted for 95 per cent of all imports.

Just under 2 per cent of the reported trade between 1998 and 2002 was in wild-collected plants. The United States, the Russian Federation and China were the main suppliers of wild material with Germany, Japan and the United Kingdom being the main importers of wild-collected plants.

[Note to speaker: CITES trade data can be downloaded from the UNEP-WCMC CITES Trade Database. This can be accessed online via the CITES Secretariat website: www.cites.org]

Slide 17: ***Cypripedium lichiangense*** **and** ***C. palangshanense***

Cypripedium lichiangense was described in 1994 from northeast Myanmar, northwest Yunnan and southwest Sichuan in China. This is a much sought-after species with a spotted birds egg-coloured flower nestling on a pair of leaves. This species is particularly difficult to grow and propagate; plants rarely survive more than 3-4 years in cultivation.

Although *Cypripedium palangshanense* was described in 1936, this species, with small purple flowers, remained largely unknown to the western world until it was rediscovered in 1998. It occurs as a narrow endemic of northwestern and eastern Sichuan in China. This species has a very slender, creeping rhizome.

Both species have found their way into trade.

[Note to speaker: The slide shows Cypripedium lichiangense (left) and Cypripedium palangshanense (top & bottom right).]

Slide 18

Slide 18: *Phragmipedium*

Phragmipedium is a small genus, comprising some 20 species, ranging from southern Mexico and Guatemala, through Central America to South America as far as southern Bolivia and Brazil. They can be ground-growing, or grow on rock surfaces or trees for support. Typically they are found growing around waterfalls and in other damp areas.

They have a creeping rhizome (modified root-like stem) with leathery leaves that are V-shaped in cross-section. They are less popular in trade than their Southeast Asian rival, *Paphiopedilum*. However, the discovery of *Phragmipedium besseae* in the 1980s, with its novel red flowers caused a marked increase in trade and renewed hobbyist interest in this genus. More recently, in 2002, the discovery of *Phragmipedium kovachii* in Peru, with its large purple flowers, has further stimulated interest in this group.

[Note to speaker: The slide shows Phragmipedium besseae (left and top centre) and Phragmipedium longifolium (right and bottom centre).]

Slide 19: *Phragmipedium* Characteristics – Flowers

When the flowers are in bud, the petal-like modified leaves (sepals) that surround the flower touch at the margins. In the other groups, these sepals overlap. In addition, the margin of the lip of the slipper-like flower folds inward. When the flower withers, unlike *Cypripedium*, the flower is deciduous and falls from the end of the fruit.

[Note to speaker: The slide shows Phragmipedium besseae var. dalessandroi (left), Phragmipedium kovachii (left centre, top), Phragmipedium longifolium (right centre, top), Phragmipedium lindenii (right), Phragmipedium schlimii (left centre, bottom), Phragmipedium wallisii (right centre, bottom).]

Slide 20

Slide 20: *Phragmipedium* Characteristics – Vegetative

Leaves of *Phragmipedium* are oblong to linear and are usually dull or glossy green with no markings. The leaves are flat, leathery, occur in two ranks or rows on opposite sides of the stem, and have a prominent midrib but lack obvious veins. Leaves persist for two or more years and are not shed on an annual basis. Typically the leaves are long and strap-like and tend to have a pointed leaf tip. Stems are characteristically short. All species have a short or creeping rhizome which lacks the annual growth 'beads' seen in *Cypripedium.*

[Note to speaker: The slide shows Phragmipedium besseae (left), Phragmipedium lindleyanum (top centre), Phragmipedium longifolium (bottom centre) and the pointed leaf tips of a Phragmipedium species (right).]

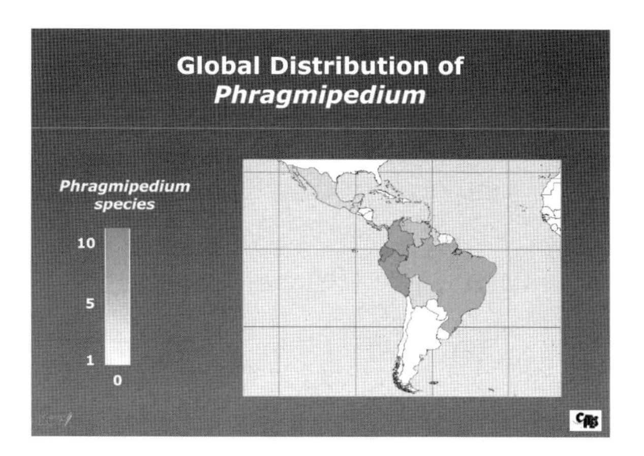

Slide 21: Global Distribution of *Phragmipedium*

Phragmipedium is found in Central and South America. It ranges from southern Mexico and Guatemala to southern Bolivia and Brazil.

Slide 22

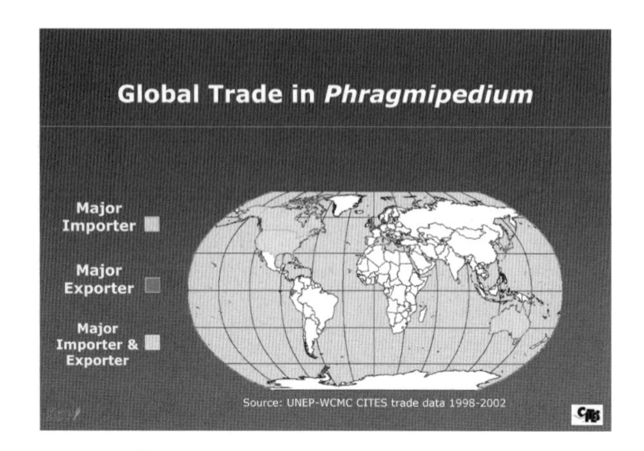

Slide 22: Global Trade in *Phragmipedium*

The recorded trade in taxa for the years 1998–2002 accounted for less than 2 per cent of all international trade in slipper orchids. All CITES recorded trade for *Phragmipedium* taxa in that period was confined to artificially propagated material.

Ecuador, the United Kingdom, Taiwan (Province of China) and the United States were the largest exporters of artificially propagated material in that period; accounting for 84 per cent of the total. Japan, the United States, Canada and Australia accounted for 77 per cent of all *Phragmipedium* imports in that period. The trade data for 1998 to 2002 also indicates that many countries received their plants as re-exports from the United States.

[Note to speaker: CITES trade data can be downloaded from the UNEP-WCMC CITES Trade Database. This can be accessed online via the CITES Secretariat website: www.cites.org]

Slide 23: *Phragmipedium kovachii*

Considerable publicity has been generated by the discovery and importation into the USA of *Phragmipedium kovachii*. It was discovered in northern Peru and described in 2002. The media interest and controversy surrounding its discovery obscured the fact that the species was thought to have been collected to extinction by collectors. However, of the 5 known sites, at least one still existed in 2004, with several hundreds of plants. The Peruvian Government have approved a very limited collection from the wild to allow controlled propagation by seed. It is reported that they are allowing export of selected species lines and hybrids from at least one approved nursery. The successful conservation of this unusual and sought-after species can only be aided by mass propagation of seedlings, as quickly as possible, to undermine the value of illegal wild collected plants.

[Note to speaker: This slide shows three images of Phragmipedium kovachii.*]*

Slide 24

Slide 24: *Paphiopedilum*

Paphiopedilum is the largest genus of slipper orchids with some 80 species occurring in the Asian tropics from southern India to New Guinea and the Philippines. They can be found growing on the ground, on rock and cliff surfaces and attached to trees and other vegetation. Most are ground-growing, growing in leaf litter or in cracks in rocks containing organic matter. They occur in a wide range of habitats from branches of large trees in the rainforests of Thailand to the harsh serpentine soils on Mt Kinabalu.

All have the characteristic slipper-like flower, some with exaggerated inflated flowers such as in the 'Bubblegum' orchid, *Paphiopedilum micranthum,* shown on the right of this slide. They are the most popular group of slipper orchids for collectors and growers. They are in the top 5 most horticulturally important orchid genera with numerous artificial hybrids being made and produced each year. However, there has been significant illegal collection and trade in species *Paphiopedilum.*

[Note to speaker: This slide shows Paphiopedilum exul (left), Paphiopedilum rothschildianum (centre) and Paphiopedilum micranthum (right).]

Slide 25: *Paphiopedilum* Characteristics – Flowers

When in bud, the petal-like modified leaves (sepals) that surround the flower overlap each other. In addition, the margin of the slipper-like flower lip does not fold over. When flowers wither, unlike *Cypripedium*, the flower is deciduous and falls from the end of the fruit. In many species the flowers bear hairs and wart-like attachments. Many species are solitary-flowered but some bear multiple flowers. Flower colour is wide, ranging through green to white, pure gold and purple. Some of the most sought-after species have flowers with long, drooping petals, reaching, in the case of *Paphiopedilum sanderianum*, over one metre in length.

[Note to speaker: The slide shows Paphiopedilum malipoense (left), Paphiopedilum haynaldianum (left centre, top), Paphiopedilum bellatulum (right centre, top), Paphiopedilum druryi (left centre, bottom), Paphiopedilum liemianum (right centre, bottom) and Paphiopedilum sanderianum (right).]

Slide 26

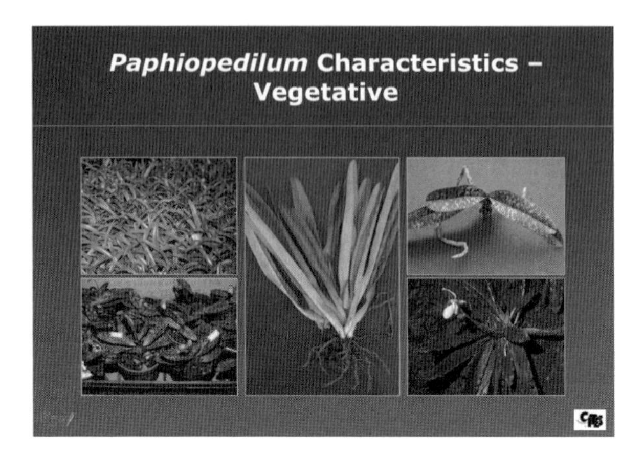

Slide 26: *Paphiopedilum* Characteristics – Vegetative

A rhizome is present in all species but it is usually short. A small number of species, *Paphiopedilum bullenianum*, *P. armeniacum*, *P. micranthum* and *P. druryi* have runner-like rhizomes up to a metre long.

Paphiopedilum leaves are leathery with a prominent middle rib. The leaves are V-shaped in cross-section. Leaves may be short and strap-like or oblong to linear. The leaves are usually short, less than 20 centimetres in length. An exception to this rule is the multiflowered group which includes species such as *Paphiopedilum sanderianum*, *P. rothschildianum* and *P. lowii.* Leaf colour ranges from plain or glossy green to mottled purple and can be quite useful in identification.

[Note to speaker: The slide shows plants with leathery green strap shaped-leaves (top left), mottled leaves (bottom left), leaves longer than 20 cm in length (centre), mottled leaves (top right) and in habitat with glossy green strap-shaped leaves (Paphiopedilum liemianum bottom right).]

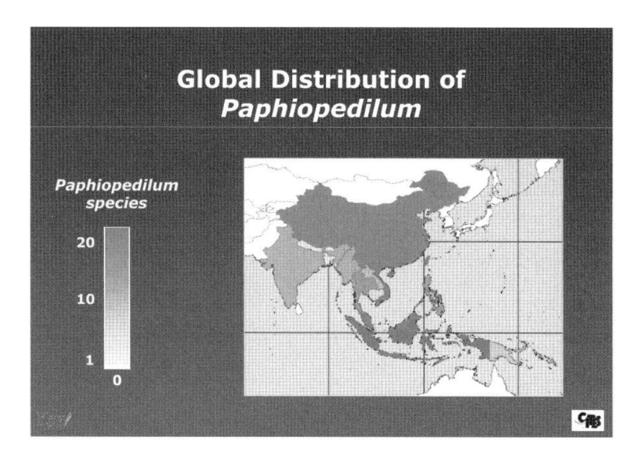

Slide 27: Global Distribution of *Paphiopedilum*

The genus *Paphiopedilum* includes some 80 species. Many of these are narrow endemics, but the genus as a whole occurs throughout Southeast Asia, extending from India, east across China to the Philippines and south through the Malay Archipelago to New Guinea and the Solomon Islands. Currently China and Viet Nam are home to many of the most sought-after species.

Slide 28

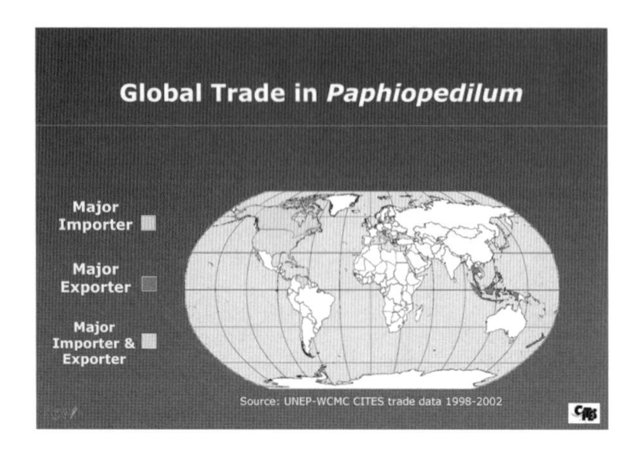

Slide 28: Global Trade in *Paphiopedilum*

A wide range of countries reported exporting artificially propagated *Paphiopedilum* taxa between 1998 and 2002. Indonesia, the Netherlands, Thailand and New Zealand were the major exporting states, accounting for 79 per cent of all recorded exports. In addition, Taiwan (Province of China), the United States, Japan and Belgium all exported more than 10,000 plants. Together these 8 exporters account for just over 90 per cent of all exports. Japan, Malaysia and Austria bring that figure to just over 97 per cent of all exports.

The two largest importers between 1998 and 2002 were Japan and the United States, between them accounting for 75 per cent of all recorded imports. Together with Canada, Italy, Switzerland, Hong Kong and Venezuela, these states accounted for 93 per cent of imports in that period.

[Note to speaker: CITES trade data can be downloaded from the UNEP-WCMC CITES Trade Database. This can be accessed online via the CITES Secretariat website: www.cites.org]

Slide 29: China and Viet Nam

In the 1980s and early 1990s a new group of *Paphiopedilum* were found in China. Not only were they exciting new species, unlike any others previously found but they also represented a whole new potential breeding line for hybrids. These included *Paphiopedilum armeniacum*, *P. emersonii, P. micranthum* and *P. malipoense*. This resulted in significant illegal trade. These species are now fairly well established in cultivation, which has had the effect of reducing the need to collect wild plants, although new colour forms such as albinos are always sought after. However, the pressure of collection for international trade has made these species vulnerable to extinction in China.

Towards the end of the 1990s and early 2000s a number of new and unusual species of slipper orchids were described from Viet Nam, most notably *Paphiopedilum vietnamense* and *P. hangianum*. Both are highly sought-after, due to their novel forms. *Paphiopedilum vietnamense* was first described in 1999. In 2001, an expedition was launched to survey the only known locality and only a handful of seedlings were located. This species is now considered to be Critically Endangered because of its very restricted range and level of exploitation.

[Note to speaker: The slide shows Paphiopedilum armeniacum (left), Paphiopedilum emersonii (top centre), Paphiopedilum malipoense (bottom centre) and Paphiopedilum micranthum (right).]

Slide 30

Slide 30: Slipper Orchids on CITES: Summary

In this section we have outlined:

- the three genera of slipper orchids covered by CITES: *Cypripedium*, *Phragmipedium* and *Paphiopedilum;*
- their characteristics, global distribution and trade.

[Note to speaker: The slide shows: Phragmipedium wallisii (left), Cypripedium flavum (centre) and Paphiopedilum helenae (right).]

Implementing CITES
for Slipper Orchids

Slide 32

Slide 32: Enforcement

The enforcement of CITES controls is carried out at different levels. Within an exporting country it is carried out by the inspection of nurseries, traders, markets and, less frequently but most importantly, of the plants at time of export. Inspections can also occur at the time of import and post-importation in the major trading countries. Enforcement agencies also survey trade shows, advertisements in the trade press and the World Wide Web.

Few countries have enforcement teams that are specially trained to identify CITES specimens - animals or plants. CITES enforcement for plants is most likely to be carried out by general Customs staff or by officials trained in plant health controls. When CITES enforcement is carried out by general Customs staff the enforcement procedures are concentrated on the documentation, not the plants. Thus, Customs may check to see if the permits are correctly filled in, stamped and issued by the correct authorities. They also check other documents and invoices to see if any CITES material named on the accompanying documentation is missing from the CITES permits.

Where general Customs staff are used to check CITES plants, it is vital that they have access to a centre of expertise on the identification and conservation of plants. Such a centre should be the national Scientific Authority. However, in some cases the national Scientific Authority may be a committee or a government department with expertise centred on animals. In this case, the enforcement authorities should build a relationship with a national or local botanic garden or herbarium. Such a relationship is vital.

The Customs officers will need some basic training on the plants and parts and derivatives covered by CITES and will need help on targeting detrimental trade. Most importantly, Customs officials will need access to experts who can identify CITES plants. Such experts can also advise on, and have access to facilities for holding, seized or confiscated material. These scientists may be called on to be expert witnesses, who are vital if breaches of the controls result in prosecution and court appearances.

Slide 32

[Note to speaker: To identify the relevant CITES Secretariat staff member to contact on enforcement issues check the staff list on the CITES website: www.cites.org.]

Slide 33

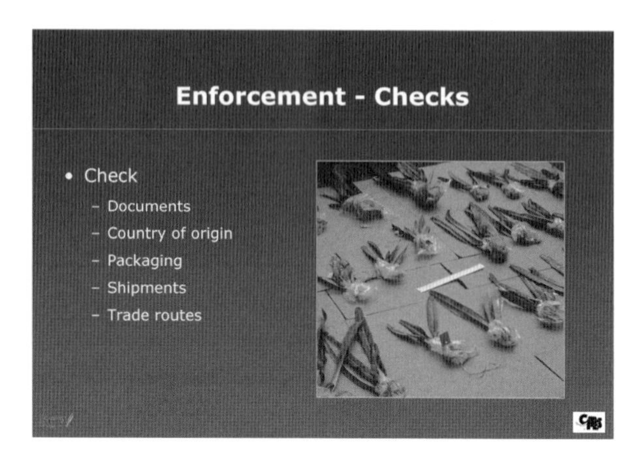

Slide 33: Enforcement – Checks

Documents - Check the authenticity of the CITES permits (signatures, stamps), and check the plant names and number of specimens on the permit against the delivery note or invoice. Also, check the source of the plants - are they declared as wild or artificially propagated? Is the plant a recently described species? Make use of the databases and the checklists recommended in the references and resources section. Is the material flasked seedlings or tissue cultures claimed to be exempt from CITES? If this is the case and the species are newly described you may wish to ask your Management Authority to confirm legal origin of the parent stock.

Country of origin – Always check the country of origin on the permits. Are the orchids being exported from a country where the plants grow in the wild? If so then the plants may be more likely to be wild-collected. Countries may express concern over the illegal export of their wild-collected slipper orchids and ask for the assistance of other CITES parties and non-Parties to control this trade. Normally, such a request is published as a Notification to the CITES Parties (you can find this on the CITES website: www.cites.org). Viet Nam, for example, is a country that has expressed concern at illegal international trade in native species of *Paphiopedilum*.

Packaging - Nurseries will usually wrap and package their plants carefully to avoid damaging them. They are then shipped in boxes marked with the nursery's name and with printed labels. Consignments of illegally collected plants may be poorly wrapped using local materials, contain handwritten labels (sometimes with collecting data), and the plants may not be identified to species level to disguise the fact that new unnamed species may have been collected.

Consignments of plants - Collections of illegal plants usually consist of small samples of plants of different size and age groups that are not uniform in shape. They may be damaged (broken or snapped roots), and soil and weeds or native plants may be present amongst the stems and roots. Artificially propagated plants will be uniform in size and shape, and be clean of soil, pests and diseases, weeds or native plants.

Trade routes & smuggling - Illegal collections of rare or new species may be shipped using postal or courier services or hand luggage to avoid detection.

Slide 33

Collections may also be split up and sent in several different packages to ensure both a high level of survival and that at least some of the plants will evade detection.

Slide 34

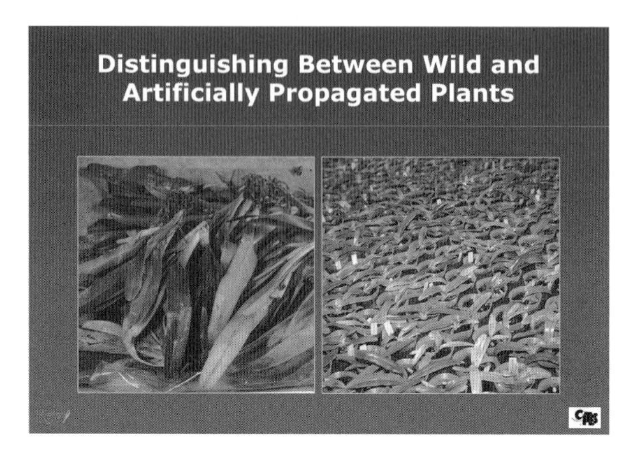

Slide 34: Distinguishing Between Wild and Artificially Propagated Plants

Distinguishing between wild-collected and artificially propagated plants is not a straightforward matter, but there are certain characteristics that can be used to make this distinction.

Wild-collected plants carry the marks of growing in their natural habitat. Plants propagated in nurseries bear the marks of an artificial, well maintained environment. They are clean, uniform and packed to a high standard. Sometimes orchids are propagated outside or in shade houses; in these cases the plants may carry some marks similar to wild collected plants. It is therefore important that you call in an expert to check the status of any plants you consider may be wild-collected rather than artificially propagated.

The CITES Identification Manual Volume 1, flora, available from the CITES Secretariat, includes details of how to distinguish wild from artificially propagated plants in the major CITES groups. But remember: <u>always get your opinion checked by an expert</u>!

Slide 35: Wild-Collected Orchids

Roots of wild-collected plants are often dead, roughly broken, or cut off in an effort to clean up the plant after collection. New roots may be growing from old damaged root material. Roots from wild plants may also have material from the natural substrate still attached. Note if the substrate attached to the roots is in any order. For example, the roots may have some organic material directly attached to them, then there may be some *Sphagnum* moss used for transport and then finally there may be some horticultural compost such as bark or rockwool. But remember; always be cautious in your assessment.

Leaves of wild-collected plants display the marks of their natural habitat, the damage caused by collection, and often the contrasting fresh growth which has occurred after collection. The base leaves are often dead or damaged. The leaves may be pitted due to desiccation and also carry the tracks made by burrowing insects. Freshly collected plants may also have growths of lichens or liverworts. Such growths would not normally survive in the controlled conditions of an orchid nursery. As the wild plants grow older after they are brought into a nursery new leaves will sprout and these will be clean and fresh in marked contrast to the old 'wild' leaves. Old 'wild' marked leaves may have been deliberately cut off, to leave only the few new leaves produced while the plant was in the nursery.

The CITES Identification Manual Volume 1, flora includes detailed information on how to tell wild-collected orchids from artificially propagated plants. However, it is always important to get an expert second opinion to confirm your identification of the plants as wild-collected. Plants grown in poor conditions outside or in shade houses sometimes carry some of the marks of wild-collected plants.

[Note to speaker: The Plants Committee has produced a series of regional directories which include contact names of CITES experts in the different countries (see CITES website for details). You may wish to use this to establish contact with a relevant expert. The characteristics of wild-collected and artificially propagated orchids are outlined in the CITES Identification Manual Volume 1, flora. Each CITES Authority is supplied with a copy of this manual by the CITES Secretariat. If your Authority does not have an up-to-date copy of this manual, contact the CITES Secretariat.]

Slide 36

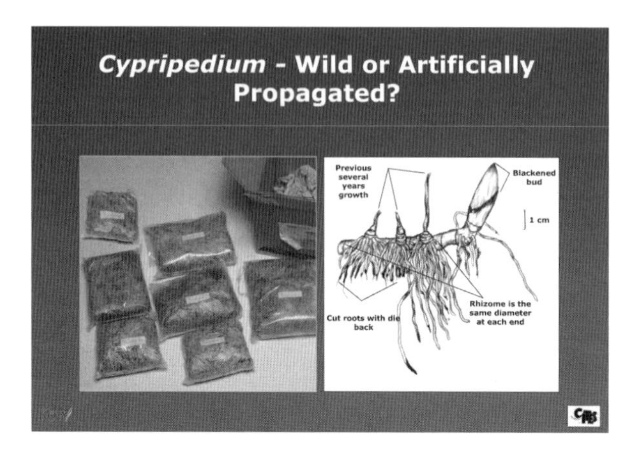

Slide 36: *Cypripedium* – Wild or Artificially Propagated?

Distinguishing between wild-collected and artificially propagated plants is not a straightforward matter, but there are certain characteristics that can help make this distinction.

All *Cypripedium* have a characteristic underground root-like modified stem called a rhizome.

The rhizome produces a series of bead-like annual growths. Most species have a short but stout, seldom-branching creeping rhizome. In some species such as *Cypripedium guttatum* and *C. margaritaceum*, the rhizome is elongated and the annual growths occur every few centimetres. The rhizome survives the dormant period with the new bud at the tip. The true roots are fibrous and emerge from behind the shoot.

Cypripedium are usually traded when dormant. They are traded, in the spring or autumn, as rhizomes displaying buds and fibrous roots. A large consignment may appear to be just a bag of roots and look nothing like what the general public considers to be a typical orchid. High quality commercial consignments are often packed in *Sphagnum* moss as in the photograph on the left.

The rhizome is extremely useful as it may be used to give an indication of the age of the plant. Round growth scars remain from the previous years growth and therefore a plant can be given a minimum age. If the rhizome is constant in diameter, this indicates a mature plant (normally at least five years old). In immature plants the rhizome display a gradual increase in diameter until it reaches its optimum size.

If you inspect a shipment and consider it to be misdeclared as artificially propagated then you should contact an expert to confirm your opinion.

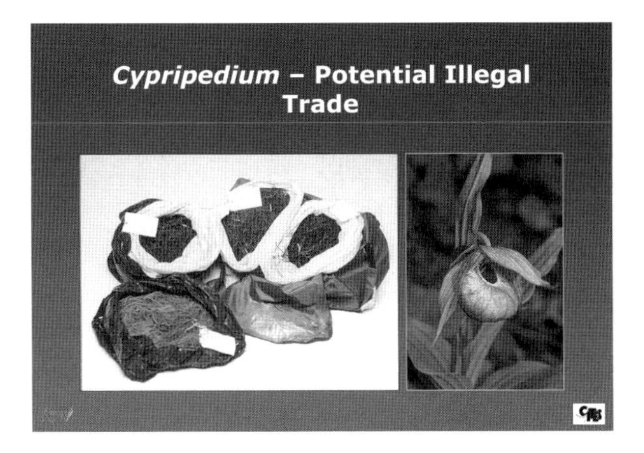

Slide 37: *Cypripedium* – Potential Illegal Trade

Illegal trade in *Cypripedium* orchids is most likely to occur in the most recently described species and those which are difficult to propagate. They may enter trade as claimed artificially propagated plants. As they are most frequently traded internationally as rhizomes, expert opinion will be required to determine if they are wild origin or artificially propagated as defined by CITES. To help you keep track of the recently described species we suggest you check the databases we have outlined in the references and resources section. Check the date that the plant was originally described to science: if it is recent, the plant is more likely to be wild-collected.

[Note to speaker: The slide shows bags of wild collected Cypripedium rhizomes (left) and Cypripedium x froschii (right).]

Slide 38

Paphiopedilum & Phragmipedium
Wild or Artificially Propagated?

	Wild	Artificially propagated
General Appearance	• Irregular shape and size • Damaged and marked	• Uniform shape and size • Clean and healthy
Roots	• Damaged or cut • Some new root growth	• Clean and healthy • May have shape of pot
Leaves	• Lower leaves damaged or cut • Insect damage/mining burrows • Pitted due to desiccation • Presence of lichens	• Clean and undamaged • Not cut back • Little or no insect damage
Soil	• Habitat soil or substrate attached	• Only horticultural compost

Slide 38: *Paphiopedilum* and *Phragmipedium* – Wild or Artificially Propagated?

This slide summarises the key characteristics of wild and artificially propagated *Paphiopedilum* and *Phragmipedium*. Orchids are usually traded internationally in a non-flowering state and initial identification and determination that the plant is wild-collected will have to be based on the examination of vegetative material. Initially, you should not be too concerned about what species the orchid is – your primary concern is to determine whether the plant is wild-collected. If the plant bears a significant number of the characteristics outlined in this slide and you think it may be wild-collected, you should contact an expert to confirm your view.

[Note to speaker: The Plants Committee has produced a series of regional directories which include contact names of CITES experts in the different countries (see CITES website for details). You may wish to use this to establish contact with a relevant expert. The characteristics of wild-collected and artificially propagated orchids are outlined in the CITES Identification Manual Volume 1, flora. Each CITES Authority is supplied with a copy of this manual by the CITES Secretariat. If your Authority does not have an up to date copy of this manual then contact the CITES Secretariat.]

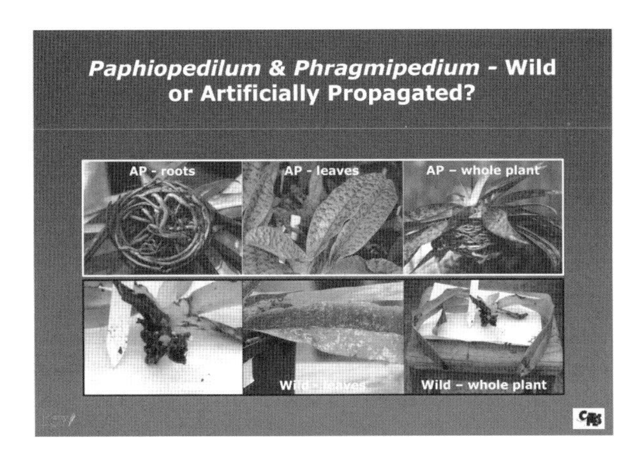

Slide 39: *Paphiopedilum* and *Phragmipedium* – Wild or Artificially Propagated?

This slide illustrates some of the characteristics that may be displayed by wild-collected and artificially propagated plants.

[Note to speaker: This slide shows Paphiopedilum spp.]

Slide 40

Slide 40: *Phragmipedium* – Potential Illegal Trade

Potential illegal trade will be concentrated on newly described species. We have highlighted the case of *Phragmipedium kovachii,* where its discovery created great interest in the orchid world and also was a spur for illegal trade. This species remains highly sought-after and wild plants may be found in illegal trade for some time. The species is likely to be smuggled without permits or misdeclared on permits.

To help you check the names of recently described species of *Phragmipedium* we suggest you consult the databases we have outlined in the references and resources section.

[Note to speaker: This slide shows three views of Phragmipedium kovachii.*]*

Slide 41: *Paphiopedilum* – Potential Illegal Trade

Once again, potential illegal trade will be concentrated on newly described species. Recently, the most interesting new species have been found in China and Viet Nam. New species have also been found in the Philippines, Indonesia and Malaysia.

Imports from Asia should be checked. In addition to commercial consignments, personal imports carried in luggage, hand baggage and sent by post and courier are all possible sources of smuggled material.

To check the names of recently described species of *Paphiopedilum* we suggest you consult the databases we have outlined in the references and resources section. You can check the date of publication of a new species name – the more recently the name has been published, the more likely the plant may be of wild origin.

[Note to speaker: This slide shows Paphiopedilum micranthum (left) and Paphiopedilum armeniacum (right) in their native habitat in China – species that were in demand from the wild when first introduced to trade in the 1980s.]

Slide 42

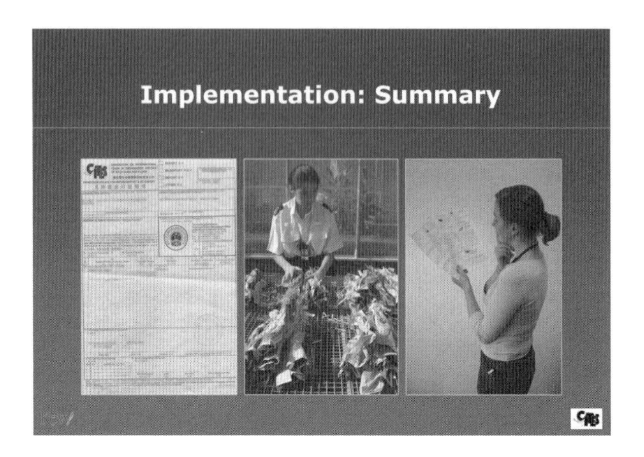

Slide 42: Implementation: Summary

We have covered the following key issues in the implementation of CITES for slipper orchids:

- enforcement procedures in different countries;
- an inspection checklist;
- the general characteristics of wild and artificially propagated plants;
- potential illegal trade.

For more information on enforcement issues and training, check the CITES website at www.cites.org.

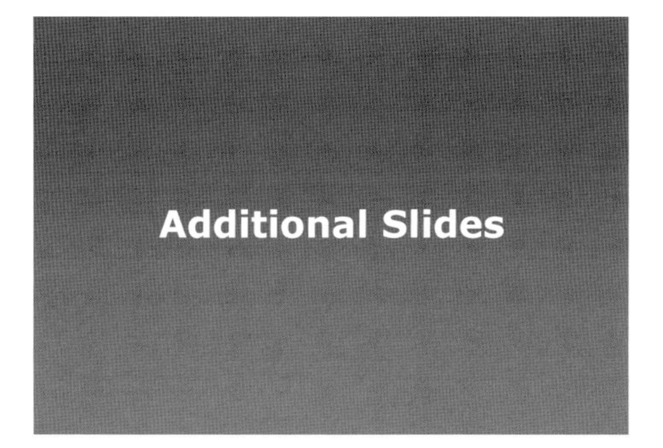
Additional Slides

Slide 44

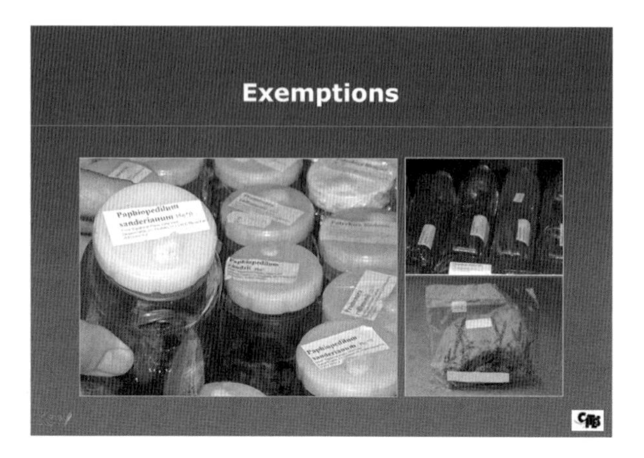

Slide 44: Exemptions

Many plants are only traded from artificially propagated sources. Recognising this, the Parties have taken the decision to exempt some orchid material from CITES controls.

For plants listed on CITES Appendix I, the whole plant and all parts and derivatives are controlled – alive or dead. There is only one exception. Seedling or tissue cultures 'obtained *in vitro,* in solid or liquid media, transported in sterile containers' are exempt. This material, of course, must be of legal origin to fulfil this exemption.

For plants listed on CITES Appendix II, the plant is controlled 'alive or dead' and so is any readily recognisable part or derivative specified in the Appendices. In the case of *Cypripedium,* the only parts and derivatives exempted from CITES control are: a) seeds and pollen (including pollinia); b) seedling or tissue cultures obtained *in vitro,* in solid or liquid media, transported in sterile containers; and c) cut flowers of artificially propagated plants.

[Note to speaker: Although a large number of orchid hybrid taxa have been exempted from CITES control, these exemptions do not extend to slipper orchids. For detail of the exemptions that apply to other orchid hybrids, check the CITES website at www.cites.org.]

Slide 45: The CITES Nursery Registration System

The CITES procedures for nursery registration are laid down in Resolution Conf. 9.19 (Rev. CoP13) *Guidelines for the registration of nurseries exporting artificially propagated specimens of Appendix I species*. This was adopted at the 9th Meeting of the CITES Conference of the Parties in Fort Lauderdale, USA, in November 1994 and revised at CoP13 in Bangkok in 2004. CITES has not laid down any criteria for the registration of nurseries that propagate Appendix II plants. However, any national CITES authority is free to set up an Appendix II registration scheme with, for example, a fast stream permit system. This would be to the benefit of the local authorities and traders, however, the registration scheme would have no recognition outside that country.

The Management Authority (MA) of any Party, in consultation with the Scientific Authority (SA), may submit a nursery for inclusion in the CITES Secretariat's Appendix I register. The owner of the nursery must first submit a profile of the operation to the MA. This profile should include, *inter alia*, a description of facilities, propagation history and plans, numbers and type of Appendix I parental stock held and evidence of legal acquisition. The MA in consultation with the SA must review this information and judge whether the operation is suitable for registration. During this process it would be normal for the national CITES authorities to inspect the nursery in some detail.

When the national authorities are satisfied that the nursery is *bona fide* and suitable for registration they pass on this opinion and the nursery details to the CITES Secretariat. The CITES MA must also outline details of the inspection procedures that they used to confirm identity and legal origin of parental stock of the plants to be included in the registration scheme and any other Appendix I material held. The national CITES authorities must also ensure that any wild-origin parental stock is not depleted and the overall operation is closely monitored. If wild-collected seeds are being used the MA should certify that the conditions outlined in the CITES definition of artificial propagation are being adhered to (see Slide 46). The MA should also put in place a fast stream permit system for the nurseries and inform the Secretariat of its details.

Slide 45

The CITES Secretariat, if satisfied with the information supplied, must then include the nursery in its register of operations. If not satisfied the Secretariat must make its concerns known to the CITES MA, indicating what needs to be clarified. Any CITES MA or other source may inform the Secretariat of breaches of the requirements for registration. If these concerns are upheld, then following consultation with the CITES MA, the nursery may be deleted from the register.

Slide 46: The CITES Definition of Artificially Propagated

The CITES definition of artificially propagated is included in Resolution Conf. 11.11 (Rev. CoP13) – *Regulation of trade in plants*. The definition within CITES includes several unique criteria. The application of these criteria may result in a plant which bears all the physical characteristics of artificial propagation being considered as wild-collected in CITES terms. The key points are:

- Plants must be grown in controlled conditions. This means, for example, the plants are manipulated in a non-natural environment to promote prime growing conditions and to exclude predators. A traditional nursery or simple greenhouse is 'controlled conditions'. A managed tropical shade house would also be an example of 'controlled conditions'. Temporary annexation of a piece of natural vegetation where wild specimens of the plants already occur would not be 'controlled conditions'. Also, wild-collected plants are considered wild even if they have been cultivated in controlled conditions for some time.

- The cultivated parent stock must have been *established in a manner not detrimental to the survival of the species in the wild* and managed in a manner which *ensures long term maintenance of the cultivated stock.*

- The cultivated parental stock must have been *established in accordance with the provisions of CITES and relevant national laws*. This means that the stock must be obtained legally in CITES terms and also in terms of any national laws in the country of origin. For example, a plant may have been illegally collected within a country of origin, then cultivated in a local nursery and then its progeny exported, declared as artificially propagated. The progeny cannot be considered to be artificially propagated in CITES terms due to the illegal collection of the parent plants.

- Seeds can only be considered artificially propagated if they are taken from plants which themselves fulfil the CITES definition of artificially propagated. The term <u>*cultivated*</u> *parental stock* is used in order to allow some addition of fresh wild-collected plants to the parental stock. It is acknowledged that parental stock may need to be occasionally supplemented from the wild. As long as this is done in a legal and sustainable fashion, it is allowed.

- Plants and seeds may be considered artificially propagated if grown from wild-collected seeds, within a range State, if this is approved by the Management and Scientific Authorities of that country.

Applying the CITES definition is a complex mixture of checking legal origin, propagation status and non-detrimental collection. To achieve this, the assessment needs to be carried out in close co-operation between the CITES Management and Scientific Authorities. The implementation of the criteria, on a day-to-day basis, needs to be tailored to the situation in an individual CITES Party. National CITES authorities should consider producing a checklist as a means of standardising the process and informing the local plant traders.

Slide 47: Promoting Sustainable Trade and Access to Breeding Material

The listing of groups of plants and animals on CITES Appendix I in effect bans trade in wild specimens for commercial purposes. The purpose of the listing is to protect such plants and animals from detrimental trade that might drive them to extinction. An Appendix I listing should not be seen as a conservation success in itself. Rather, a conservation success is achieved when that taxon can be downlisted to Appendix II. It is therefore important that conservation action take place following a CITES Appendix I listing. The demand for Appendix I taxa does not disappear following an Appendix I listing; the theory is that propagated material should be made available to fill this demand. This is possible where suitable propagation techniques have been developed and legal mother plants are available.

However, in the case of slipper orchids, new species are sought after, found and described. Legal mother plants for breeding stock are often difficult to obtain. In these cases illegal material leaks into the international market and is slowly incorporated into breeding stock. This process promotes unsustainable collection of the rarest species and robs the country of origin of the significant income that could be derived from the introduction of such stock into the international market place. This problem has not been addressed. There has been inertia within countries of origin, the international orchid trade and the CITES community. We await successful collaboration to establish mechanisms to allow access to breeding material and help defeat the illegal trade. Countries of origin need assistance in setting up such programmes.

It is possible to establish initiatives within CITES to make this happen. All that is needed is enthusiasm, initiative, trust and funding. Trust is probably the hardest of these to secure. Such breeding initiatives will always be vulnerable to being undermined by illegal trade or indeed being branded as 'biopiracy'. If you are working within CITES or the orchid industry you should try to encourage initiatives of this type. This would promote sustainable trade and allow countries of origin to have access to the funds generated by their own resources. It may also be the only way to cement the partnerships required for the long term conservation of the species and their habit.

INDEX

CITES et les orchidées « sabot de Vénus »

Une introduction aux orchidées « sabot de Vénus » couvertes par la Convention sur le commerce international des espèces de faune et de flore sauvages menacées d'extinction

Rédaction

H. Noel McGough,

David L. Roberts, Chris Brodie et Jenny Kowalczyk

Royal Botanic Gardens, Kew

Royaume-Uni

Conseil d'administration, Royal Botanic Gardens, Kew

2006

TABLE DES MATIÈRES

INTRODUCTION

'*CITES et les orchidées « sabot de Vénus »*' est une introduction aux sabots de Vénus contrôlés par la CITES. Cet ouvrage aborde l'identification et le commerce de ces plantes, ainsi que la mise en œuvre de la CITES pour ces taxons.

Ce dossier vise principalement à servir d'outil pour la formation de tous ceux qui travaillent avec la Convention, notamment les organes de gestion et les autorités scientifiques CITES ainsi que les organes chargés du respect de la Convention. Toutefois, nous espérons que son contenu pourra être utile à un public plus large, surtout celui qui s'intéresse au fonctionnement de la CITES en ce qui concerne ce groupe de plantes, de grande importance pour le commerce.

Cet ouvrage a été conçu pour être flexible et s'adapter facilement aux besoins de chaque utilisateur. Nous encourageons les utilisateurs à développer leur exposé en fonction des besoins et intérêts de leur public. Outre les notes explicatives des diapositives, nous avons ajouté une bibliographie et une liste de ressources. Nous espérons que ce dossier sera un instrument utile pour réaliser vos propres présentations, et qu'il servira d'ouvrage de référence pratique. Merci d'utiliser cet outil de formation et de nous adresser vos commentaires afin que nous puissions réviser les éditions futures en les adaptant à vos besoins.

Noel McGough

Responsable de la Section des conventions et des politiques

Autorité scientifique CITES du Royaume-Uni chargée des plantes

Royal Botanic Gardens, Kew

REMERCIEMENTS

Les auteurs remercient les personnes suivantes pour leur savoir-faire et leur assistance technique : Wendy Byrnes, Phillip Cribb, Margarita Clemente Muñoz, Deborah Rhoads Lyon, Matthew Smith, Sabina Michnowicz et Ger van Vliet.

Le dossier a été financé par l'organe de gestion CITES du Royaume-Uni, le Département pour l'environnement, l'alimentation et les affaires rurales (Defra).

COMMENT UTILISER CE DOSSIER ?

Ce dossier contient des diapositives accompagnées de notes explicatives conçues pour présenter les orchidées « sabot de Vénus » inscrites aux annexes de la CITES. L'exposé se divisera en trois axes thématiques indépendants pouvant être utilisés et adaptés aux caractéristiques, intérêts et besoins de votre public (Introduction aux sabots de Vénus, Sabots de Vénus couverts par la CITES et Mise en œuvre de la CITES pour les sabots de Vénus).

Une quatrième partie avec des diapositives supplémentaires et des notes pour l'orateur contient des informations détaillées sur d'autres thèmes que vous pourrez ajouter à votre présentation dans la mesure où vous le jugerez utile. Les diapositives ont été rédigées d'une manière générale dans l'intention de rester à jour et de ce fait utiles à moyen terme.

Chaque diapositive est accompagnée de notes explicatives conçues pour guider l'orateur. Les notes sont plus détaillées que les diapositives et contiennent des informations mises à jour en mai 2005. Naturellement, nous encourageons chaque orateur à utiliser son style personnel et à se rapprocher ou s'éloigner des notes selon ses propres critères.

Nous espérons que ce dossier sera un point de départ utile à partir duquel les diapositives et les notes seront adaptées en fonction des besoins particuliers de votre public, de la durée de l'exposé et de votre style personnel. Par exemple, vous pouvez illustrer quelques diapositives par des exemples pris dans votre région ou votre institution, ou même ajouter des images (bandes dessinées, photos, extraits de presse, etc.). Ces détails augmenteront certainement la portée d'une présentation individuelle. De plus, les diapositives peuvent être imprimées sur des transparents et projetées sur un écran. Les cas échéant, elles peuvent être polycopiées à partir du dossier ou bien imprimées à partir du fichier Microsoft® PowerPoint® du CD-ROM pour être distribuées au public.

CD-ROM

Le CD-ROM contient les fichiers suivants:

- 'CITESSlipperOrchids.ppt', une présentation Microsoft PowerPoint® qui comprend les diapositives et les notes explicatives. Vous aurez besoin de Microsoft PowerPoint 97® (ou une version plus récente) pour visualiser et personnaliser ce fichier.

- 'CITESSlipperOrchids.pdf', une présentation en format Adobe Acrobat®. Vous ne pourrez pas modifier cette présentation, mais elle peut être visualisée en plein écran en utilisant Adobe Acrobat Reader®. Pour télécharger Adobe Acrobat Reader®, (visitez www.adobe.com).

- 'CITESSlipperOrchidsBW.pdf,' une présentation en noir et blanc en format de Adobe Acrobat®.

- 'CITESSlipperOrchidsPack.pdf', une copie complète du texte du dossier y compris l'introduction, les références et notes explicatives. Ce fichier permet de visualiser le document électronique complet ainsi que d'imprimer le dossier entièrement ou en partie. Vous aurez besoin de Adobe Acrobat Reader® pour visualiser ce fichier.

REFERENCES ET RESSOURCES

Références sur la Convention

CITES (2003 et versions mises à jour). *Guide CITES.* Secrétariat de la Convention sur le commerce international des espèces de faune et de flore sauvages menacées d'extinction. Genève, Suisse. Ce guide comprend le texte de la Convention et ses annexes, le modèle standard de permis ainsi que le texte des résolutions et des décisions de la Conférence des Parties.

Wijnstekers, W. (2003 et versions mises à jour). *The Evolution of CITES, 6th edition.* Secrétariat de la Convention sur le commerce international des espèces de faune et de flore sauvages menacées d'extinction. Genève, Suisse. La référence la plus complète et prépondérante sur la Convention dont nous disposons. Ouvrage écrit par le Secrétaire général de la CITES, et mis à jour régulièrement.

Rosser, A. et Haywood, M. (Compilateurs), (2002). *Guidance for CITES Scientific Authorities. Checklist to assist in making non-detriment findings of Appendix II exports.* Occasional Paper of the IUCN Species Survival Commission No. 27. IUCN - The World Conservation Union, Gland, Suisse et Cambridge, Royaume-Uni. La première tentative de définir les lignes directrices à suivre par les Autorités scientifiques lorsqu'elles émettent leurs avis de commerce non préjudiciable requis avant la délivrance des permis d'exportation CITES.

Le site Web de la CITES (www.cites.org) contient une grande quantité d'information sur la Convention, les espèces inscrites aux annexes, les adresses et contacts clés, les rapports issus des réunions et des groupes de travail, les nouvelles publications et les sites Web, ainsi qu'un calendrier d'évènements.

Analyse critique de la Convention

Hutton, J. et Dickson, B. (2000). *Endangered Species, Threatened Convention. The Past, Present and Future of CITES.* Earthscan, Londres, Royaume-Uni. Une évaluation critique de la CITES du point de vue de l'utilisation durable.

Oldfield, S. (Editeur), (2003). *The Trade in Wildlife: Regulation for Conservation.* Earthscan, Londres, Royaume-Uni. Une évaluation critique du commerce international des espèces sauvages. Reeve, R. (2002*). Policing International Trade in Endangered Species. The CITES Treaty and Compliance.* Royal Institute of International Affairs. Earthscan. Londres, Royaume-Uni. Une étude détaillée du système de mise en application et de lutte contre la fraude de la CITES.

Références normalisées CITES pour les plantes – Listes de référence

Carter, S. et Eggli, U. (2003). *The CITES Checklist of Succulent Euphorbia Taxa (Euphorbiaceae).* Second edition. German Federal Agency for Nature Conservation, Bonn, Allemagne. Référence pour les noms d'espèces succulentes des *Euphorbia.*

Hunt, D. (1999). *CITES Cactaceae Checklist.* Second edition. Royal Botanic Gardens, Kew, Royaume-Uni. Référence pour les noms des cactacées.

Mabberley, D.J. (1997). *The Plant-Book.* Second edition. Cambridge University Press, Cambridge, Royaume-Uni. La référence pour les noms génériques de toutes les espèces végétales inscrites aux annexes de la CITES, à moins qu'ils ne soient remplacés par les listes de références normalisées adoptées par les Parties et mentionnées sur cette liste.

Newton, L.E. et Rowley, G.D. (Eggli, U. Editeur), (2001). *CITES Aloe and Pachypodium Checklist.* Royal Botanic Gardens, Kew, Royaume-Uni. Référence pour les noms des genres *Aloe* et *Pachypodium.*

Roberts, J.A., Beale, C.R., Benseler, J.C., McGough, H.N. et Zappi, D.C. (1995). *CITES Orchid Checklist. Volume 1.* Royal Botanic Gardens, Kew, Royaume-Uni. Référence pour les noms de *Cattleya, Cypripedium, Laelia, Paphiopedilum, Phalaenopsis, Phragmipedium, Pleione* et *Sophronitis.* Comprend des descriptions de *Constantia, Paraphalaenopsis* et *Sophronitella.*

Roberts, J.A., Allman, L.R., Beale, C.R., Butter, R.W., Crook, K.B. et McGough, H.N. (1997). *CITES Orchid Checklist. Volume 2.* Royal Botanic Gardens, Kew, Royaume-Uni. Référence pour les noms de *Cymbidium, Dendrobium, Disa, Dracula* et *Encyclia.*

Roberts, J.A., Anuku, A., Burdon, J. , Mathew, P., McGough, H.N. et Newman, A.D. (2001). *CITES Orchid Checklist. Volume 3.* Royal Botanic Gardens, Kew, Royaume-Uni. Référence pour les noms de *Aerangis, Angraecum, Ascocentrum, Bletilla, Brassavola, Calanthe, Catasetum, Miltonia, Miltonioides, Miltoniopsis, Renanthera, Renantherella, Rhynchostylis, Rossioglossum, Vanda* et *Vandopsis.*

Willis, J.C., révisé par Airy Shaw, H.K. (1973). *A Dictionary of Flowering Plants and Ferns.* 8th edition. Cambridge University Press. Cambridge, Royaume-Uni. Pour les synonymes géneriques qui ne figurent pas dans *The Plant-Book*, à moins qu'ils ne soient remplacés par les listes de réference normalisées adoptées par les Parties à la CITES et mentionnées sur cette liste.

UNEP-WCMC (2005). *Checklist of CITES Species.* UNEP-WCMC, Cambridge, Royaume-Uni. La CdP a adopté cette liste de référence et ses versions mises à jour comme compilation officielle des noms scientifiques inclus dans les références normalisées.

Les listes de référence de la CITES sont mises à jour régulièrement par le Comité de la nomenclature de la Convention. Pour de plus amples informations, consultez le site Web de la CITES : www.cites.org.

Références supplémentaires

Voici quelques références générales qui, nous le souhaitons, vous seront utiles. Soyez conscient que la taxonomie qui figure dans ces ouvrages pourrait être différente de celle prescrite dans les références normalisées adoptées pour la CITES, mentionnées plus haut. Merci de nous faire part des ouvrages que vous considérez utiles afin que nous puissions les ajouter dans les éditions futures de ce guide.

Averyanov, L., Cribb, P., Loc P.K. & Hiep. N.H (2003). *Slipper Orchids of Vietnam.* Royal Botanic Gardens, Kew. Royaume-Uni. Ouvrage détaillé contenant une description complète de toutes les espèces de *Paphiopedilum* originaires du Vietnam, des dessins au trait et une grande quantité de photos couleur, y compris quelques photos des habitats.

Références et ressources

Bechtel, H. Cribb, P. et Launert, E. (1992). *The Manual of Cultivated Orchid Species*. Third Edition. Blanford Press, Londres, Royaume-Uni. Cet ouvrage a grandement besoin d'être mis à jour, mais il est toujours une excellente référence et contient une analyse détaillée de plus de 400 genres, 1 200 espèces, plus de 860 photos couleur et de nombreux dessins au trait excellents.

Braem, G. J., Baker, C.O et Baker, M.L. (1998). *The Genus Paphiopedilum: Natural History and Cultivation, Volume 1.* Botanical Publishers, Inc., Kissimmee, Floride, Etats-Unis.

Braem, G. J., Baker, C.O et Baker, M.L. (1999). *The Genus Paphiopedilum: Natural History and Cultivation, Volume 2.* Botanical Publishers, Inc., Kissimmee, Floride, Etats-Unis.

Braem, G. J. et Chiron, G.R. (2003). *Paphiopedilum.* Tropicalia, Saint-Genis Laval, France.

Cash, C. (1991). *The Slipper Orchids.* Christoper Helm, Londres, Royaume-Uni.

Cavestro, W. (2001). *Le genre Paphiopedilum: taxonomie, répartition, habitat, hybridation et culture.* Rhône-Alpes Orchidées, Lyon, France.

(CITES (1993-). *CITES Identification Manual, Volume 1 Flora.* Secrétariat de la Convention sur le commerce international des espèces de faune et de flore sauvages menacées d'extinction. Genève, Suisse. Il s'agit du manuel d'identification officiel de la CITES. Les Parties qui proposent une espèce pour son inscription aux annexes sont tenues d'élaborer des fiches pour le manuel si la proposition est acceptée. Ce manuel est un classeur à anneaux auquel de nouvelles fiches sont continuellement ajoutées. Cet ouvrage est essentiel pour toute personne qui travaille avec la CITES et les plantes.

Chen, V.Y. et Song, M. (2000). *Guide to CITES Plants in Trade.* (Chinese edition). TRAFFIC Asie de l'est.

Commission européenne (2002). *Les nouveaux règlements sur le commerce des espèces de faune et de flore sauvages, cinq ans après.* Office des publications officielles des Communautés européennes. Luxembourg. Brochure sur la réglementation de l'UE en matière de commerce des espèces sauvages.

Cribb, P. (1997). *Slipper orchids of Borneo.* Natural History Publications (Bornéo), Kota Kinabalu.

Cribb, P. (1997). *The Genus Cypripedium - A Botanical Magazine Monograph.* Publié en association avec Royal Botanic Gardens, Kew. Timber Press, Portland, Etats-Unis. Monographie détaillée contenant une description complète de toutes les espèces, des photos couleur et de bonnes illustrations en noir et blanc et en couleur.

Cribb, P. (1998). *The Genus Paphiopedilum (Second Edition) - A Botanical Magazine Monograph.* Publié en association avec Royal Botanic Gardens, Kew. Natural History Publications (Bornéo), Kota Kinabalu. Monographie détaillée contenant une description complète de toutes les espèces, des photos couleur et de bonnes illustrations en noir et blanc et en couleur.

Hennessy, E. F. et Hedge, T.A. (1989). *The Slipper Orchids.* Acorn Books, Randburg, Afrique du Sud.

Gruss, O. (2003). *A Checklist of the Genus Phragmipedium.* Orchid Digest. 67[4] : 213-255.

Hilton-Taylor, C. (Compilateur), (2000-). *IUCN Red List of Threatened Species.* IUCN-The World Conservation Union. Gland, Suisse et Cambridge, Royaume-Uni. Liste officielle de l'UICN des espèces animales et végétales menacées, publiée sous forme de brochure avec CD-ROM. La liste est continuellement mise à jour et améliorée. Pour obtenir la version la plus récente, consultez le site Web de la liste rouge à l'adresse www.redlist.org.

IUCN/SSC Orchid Specialist Group. (1996). *Orchids – Status Survey and Conservation Action Plan.* IUCN, Gland, Suisse et Cambridge, Royaume-Uni.

Jenkins, M. et Oldfield, S. (1992). *Wild Plants in Trade.* TRAFFIC International, Cambridge, Royaume-Uni. Résumé de la dernière analyse complète du commerce européen de plantes soumises aux contrôles de la CITES.

Koopowitz, H. (2000). *A revised checklist of the Genus Paphiopedilum.* Orchid Digest. 64[4] : 155-179.

Lange, D. et Schippmann, U. (1999). *Checklist of medicinal and aromatic plants and their trade names covered by CITES and EU Regulation 2307/98 Version 3.0.* German Federal Agency for Nature Conservation. Bonn, Allemagne.

Marshall, N.T. (1993). *The Gardener's Guide to Plant Conservation.* TRAFFIC North America. Malheureusement, ce guide est maintenant dépassé. Cependant, il se pourrait qu'une nouvelle édition soit en cours de préparation. Il s'agissait d'un guide très utile sur les plantes commercialisées pour l'horticulture et leurs sources.

Mathew, B. (1994). *CITES Guide to Plants in Trade.* UK Department of the Environment, Londres, Royaume-Uni. Maintenant dépassé, mais contient des photos couleur et des descriptions des principaux groupes de plantes contrôlées par la CITÉS et commercialisées au début des années 1990.

McCook, L. (1998). *An annotated checklist of the genus Phragmipedium.* Orchid Digest Corp. CA, Etats-Unis. Publication spéciale de Orchid Digest.

Pridgeon, A. (2003). *The Illustrated Encyclopedia of Orchids.* David and Charles, Devon, Royaume-Uni. Plus de 1100 espèces illustrées. Cet ouvrage passe en revue les principaux taxons qui font l'objet de commerce et qui intéressent les collectionneurs. Photos entièrement en couleur. Le meilleur guide général imprimé disponible sur les orchidées.

Rittershausen, W. & B. (1999). *Orchids – a practical guide to the world's most fascinating plants.* The Royal Horticultural Society, réimpression en 2004. Quadrille Publishing Ltd, Londres, Royaume-Uni.

Sandison, M. S., Clemente Muñoz, M., de Koning J. et Sajeva, M. (1999). *La CITES et les plantes – Guide de l'utilisateur.* Royal Botanic Gardens, Kew, Royaume-Uni. Premier « dossier de diapositives » avec 40 diapositives et texte publié en anglais, français et espagnol.

Sandison, M. S., Clemente Muñoz, M., de Koning J. et Sajeva, M. (2000). *CITES and Plants - A User's Guide.* (Chinese Edition). Royal Botanic Gardens, Kew, Royaume-Uni. Edité par Vincent Y. Chen et Michael Song et réalisé par TRAFFIC Asie de l'Est. Le Guide de l'utilisateur en chinois.

Références et ressources

Schippmann, U. (2001). *Medicinal Plants Significant Trade Study CITES Project S-109. Plants Committee Document PC9 9.1.3.(rev.). BfN - Skripten 39.* German Federal Agency for Nature Conservation, Bonn, Allemagne. Excellente vue d'ensemble de la conservation et du commerce de plantes médicinales inscrites aux annexes de la CITES.

CD-ROM

CITES (2002-). *Présentations pour la formation CITES.* Secrétariat CITES, Genève, Suisse. Série de matériels de formation (présentations Microsoft PowerPoint®) réalisés par l'Unité chargée au Secrétariat CITES du renforcement des capacités. En format CD-ROM « de poche ». Ce sont des outils essentiels pour toute personne chargée de la formation en matière de la CITES.

CITES (2003-). *CD-ROM version of the CITES website* (*www.cites.org*). Version complète du site Web de la CITES sur CD-ROM. Disponible en s'adressant au Secrétariat CITES.

Sites Web

Il existe de nombreux sites Web qui peuvent intéresser les personnes qui travaillent avec la CITES. Bon nombre d'organes CITES nationaux consacrent un site Web à la Convention. Voici une liste de sites clés qui vous mèneront à d'autres sites qu'il vous sera possible de consulter.

CITES, page d'accueil : Site officiel du Secrétariat CITES. Il contient une liste des Parties, les résolutions ainsi que d'autres documents. www.cites.org.

Commission de survie des espèces de l'UICN : La CSE (SSC, en anglais) est la source principale d'information scientifique et technique de l'UICN pour la conservation des espèces animales et végétales menacées et vulnérables. Elle mène des tâches spécifiques au nom de l'UICN telles que le contrôle des espèces vulnérables et de leurs populations, ainsi que la mise en œuvre et l'examen de plans d'action pour la conservation. Elle fournit également des lignes directrices et des conseils, et recommande des politiques à suivre aux gouvernements, aux agences et aux organisations en ce qui concerne la conservation et la gestion des espèces et de leurs populations. www.iucn.org/themes/ssc/.

Commission européenne : Information sur la réglementation du commerce des espèces sauvages qui sert à mettre en œuvre la CITES au sein de l'Union européenne. www.eu-wildlifetrade.org.

Earth Negotiations Bulletin (Bulletin des Négociations de la Terre) : Ce service suit les négociations environnementales les plus importantes au fur et à mesure qu'elles se déroulent. Ce site contient également une grande quantité d'archives et de nombreuses photos des réunions. www.iisd.ca.

PNUE – WCMC (Centre mondial de surveillance continue de la conservation de la nature) (UNEP – WCMC, en anglais) : Le PNUE-WCMC fournit des services d'information sur la conservation et l'utilisation durable des ressources vivantes de la terre, et aide des tiers à développer des services d'information. L'une de ses activités est de servir de soutien au Secrétariat CITES. Pour obtenir des informations sur le commerce international des espèces sauvages et des statistiques sur le commerce, il faut s'adresser au Programme sur les espèces (Species Programme, en anglais) du PNUE-WCMC. Le travail du centre, qui est à présent un bureau des Nations Unies à Cambridge, au Royaume-Uni, fait partie intégrale du Programme des Nations Unies pour l'environnement (PNUE), dont le siège se trouve à Nairobi, au Kenya. www.unep-wcmc.org/index.html.

Site Web du Royaume-Uni sur la CITES : Site Web maintenu par les organes CITES du Royaume-Uni, et qui vise à fournir des informations à jour sur des questions relatives à la CITES qui touchent e Royaume-Uni et ses territoires d'outremer. www.ukcites.gov.uk.

TRAFFIC International : TRAFFIC est un programme de WWF et de l'UICN établi pour contrôler le commerce d'espèces végétales et animales sauvages. Le Réseau TRAFFIC est le programme de contrôle le plus important à l'échelon mondial, avec des bureaux distribués dans le monde entier. Le Réseau travaille en étroite collaboration avec le Secrétariat CITES. www.traffic.org.

UICN – Union mondiale pour la nature : La plus grande organisation professionnelle consacrée à la conservation. L'UICN rassemble des gouvernements, des organisations non gouvernementales, des institutions et des individus afin d'aider les nations à faire le meilleur usage de leurs ressources naturelles de manière durable. www.iucn.org.

Vérification des noms des plantes

Les sites Web suivants sont utiles pour vérifier les noms de plantes que l'on ne trouve pas dans les listes de référence normalisées de la CITES. Parfois il s'agit de noms d'espèces récemment décrites. Si ce « nouveau nom » a été utilisé sur une demande de permis CITES en déclarant que la plante a été reproduite artificiellement, il sera nécessaire de vérifier l'identité de la plante et de s'assurer qu'elle correspond à la définition de reproduction artificielle de la CITES.

IPNI - The International Plant Names Index : Base de données sur les noms et les détails bibliographiques de base qui correspondent à toutes les plantes à graines. www.ipni.org/index.html.

TROPICOS : Base de données en matière de nomenclature élaborée et maintenue par le Missouri Botanical Garden. mobot.mobot.org/W3T/Search/vast.html.

EPIC – Electronic Plant Information Centre : Ce site rassemble toute l'information numérisée sur les plantes détenue par Royal Botanic Gardens, Kew. www.rbgkew.org.uk/epic/.

World Checklist of Monocotyledons : Cette liste contient un inventaire des plantes monocotylédones dont la nomenclature est acceptée avec les détails bibliographiques pertinents, ainsi que leur distribution mondiale. Elle contient la liste complète de tous les noms des orchidées. www.kew.org/monocotChecklist/

Phragweb : Ce site est une source complète d'informations sur les espèces du genre *Phragmipedium* et contient des descriptions, des dessins au trait et des photos entièrement en couleur. www.Phragweb.info.

RHS : Site Web de la Royal Horticultural Society, utile pour vérifier les noms des nouveaux hybrides.
www.rhs.org.uk/publications/pubs_journals_orchid_hybrid.asp.

INDEX DES DIAPOSITIVES

Diapositives d'introduction

Introduction aux sabots de Vénus

Sabots de Vénus couverts par la CITES

Mise en œuvre de la CITES pour les sabots de Vénus

Diapositives supplémentaires

Dia 1 : La CITES et les orchidées « sabot de Vénus »

Cet exposé vise à servir d'introduction aux différents types d'orchidées « sabot de Vénus » couvertes par la Convention sur le commerce international des espèces de faune et de flore sauvages menacées d'extinction – CITES. Nous aborderons certaines des questions les plus importantes relatives à la mise en œuvre de la Convention pour cet important groupe de plantes.

Dia 2

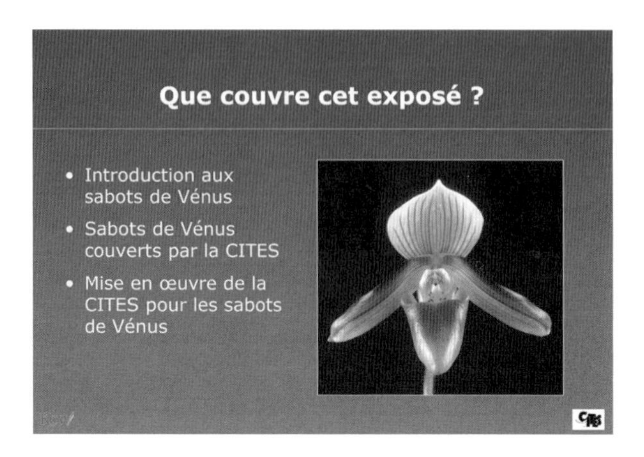

Dia 2 : Que couvre cet exposé ?

Nous aborderons les sujets suivants:

- Introduction aux sabots de Vénus ;
- Sabots de Vénus couverts par la CITES ;
- Mise en œuvre de la CITES pour les sabots de Vénus.

[Note pour l'orateur : cette diapositive montre Paphiopedilum callosum.]

Dia 3 : La diversité des orchidées

La plupart des gens ont au moins une idée générale de ce qu'est qu'une orchidée. Le nom « orchidée » évoque souvent des idées de fleurs fascinantes, de milieux exotiques et de cadeaux mystérieux. Toutefois, les plantes que remarque le grand public ne représentent qu'une fraction minuscule de celui qui est possiblement le groupe de plantes le plus large au monde. Il contient plus de 25 000 espèces répertoriées, et l'on estime que quelque 5 000 espèces restent à découvrir. Quoique ces espèces se trouvent presque partout dans le monde, la plupart – environ 70% – se concentre dans les forêts tropicales. Cependant, il est possible de trouver des orchidées même dans des régions très arides ou subartiques telles que l'Alaska. On connaît même quelques espèces souterraines qui poussent en Australie !. La taille des plantes est très variable, allant de celle de la pointe d'un crayon à celle d'un arbuste vigoureux de deux tonnes. De plus, contrairement à la croyance populaire, toutes les orchidées ne sont pas rares. Lorsqu'elles se trouvent dans un habitat approprié, certaines espèces forment de très grandes colonies. Les orchidées fascinent les collectionneurs et les amateurs de plantes depuis le XIXe siècle en raison de leur grande variété de tailles et de couleurs et de leur attrait exotique.

[Note pour l'orateur : les orchidées se distinguent des autres végétaux par un ensemble de caractéristiques florales :

1. *Les parties mâles et femelles de la fleur sont soudées entièrement ou en partie et forment une structure appelée colonne.*

2. *Un des pétales de la fleur est très modifié et constitue souvent une plate-forme d'atterrissage ou un guide pour les insectes pollinisateurs qui reçoit le nom de labelle ou lèvre.*

3. *Les grains de pollen demeurent souvent agglomérés et forment de grandes masses appelées pollinies qui sont disposées par paires ou en groupes de 4, 6 ou 8.*

4. *Les orchidées produisent des millions de graines minuscules dépourvues des éléments nutritifs nécessaires pour germer. Par conséquent, ces plantes dépendent de la présence d'un champignon qui facilite leur germination.*

Cette diapositive montre Coryanthes macrantha (à gauche), Masdevallia veitchiana (au milieu) et Dendrobium secundum (à droite).]

Dia 4

Dia 4 : Pourquoi protéger les orchidées ?

Les orchidées sont très menacées par la destruction de leur habitat et, en moindre mesure, par les prélèvements excessifs. Si la destruction de l'habitat touche toutes les espèces, les prélèvements excessifs sont particulièrement graves dans le cas des espèces d'importance commerciale et peuvent provoquer l'extinction d'une espèce dans la nature seulement quelques années après sa découverte.

L'évolution des orchidées a donné lieu à de complexes stratégies de pollinisation qui sont facilement perturbées par la destruction de l'habitat ou les prélèvements excessifs, raison pour laquelle on les considère de bons indicateurs d'environnement. Par conséquent, la conservation des orchidées et le maintien de populations adéquates de ces espèces contribue probablement à la protection de bon nombre d'autres espèces. En raison de leur popularité, les orchidées sont des espèces emblématiques qui peuvent servir à promouvoir la conservation d'habitats clés pour les espèces sauvages.

[Note pour l'orateur : la diapositive montre Phragmipedium besseae.]

Introduction aux sabots de Vénus

Dia 6

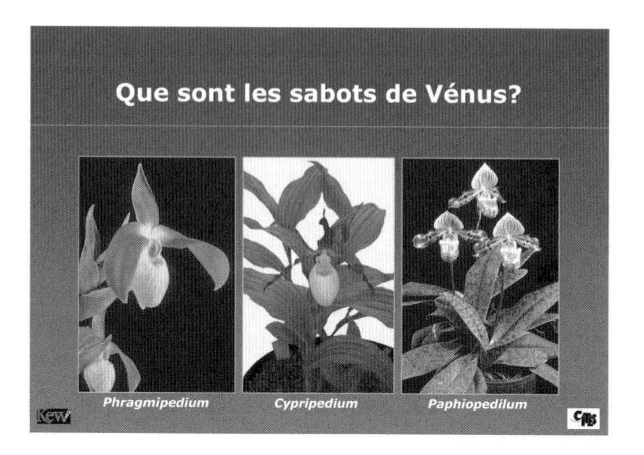

Dia 6 : Que sont les sabots de Vénus ?

Les sabots de Vénus sont faciles à distinguer des autres orchidées grâce à leurs fleurs en forme de sabot, de pantoufle ou de chaussure. La forme de ces fleurs a inspiré bon nombre de noms vernaculaires tels que « Lady's Slipper » en Europe, « Moccasin Flower » en Amérique du Nord et « Zapatilla » en Amérique latine. Les principaux genres de sabots de Vénus sont *Cypripedium*, *Paphiopedilum* et *Phragmipedium.* Tous ces genres sont en demande dans le commerce international.

Le genre *Cypripedium* comprend quelque 50 espèces réparties dans les régions tempérées de l'hémisphère Nord. Le genre *Paphiopedilum,* quant à lui, comprend quelque 80 espèces confinées en Asie du Sud-Est, et le genre *Phragmipedium* est formé par quelque 20 espèces limitées à l'Amérique centrale et l'Amérique du Sud. Ces plantes sont très attrayantes pour les collectionneurs et les horticulteurs en raison de la beauté de leurs fleurs et du nombre relativement réduit d'espèces qui existent.

[Note pour l'orateur : la diapositive montre Phragmipedium besseae var. dalessandroi (à gauche), Cypripedium parviflorum (au milieu) et Paphiopedilum venustum (à droite).]

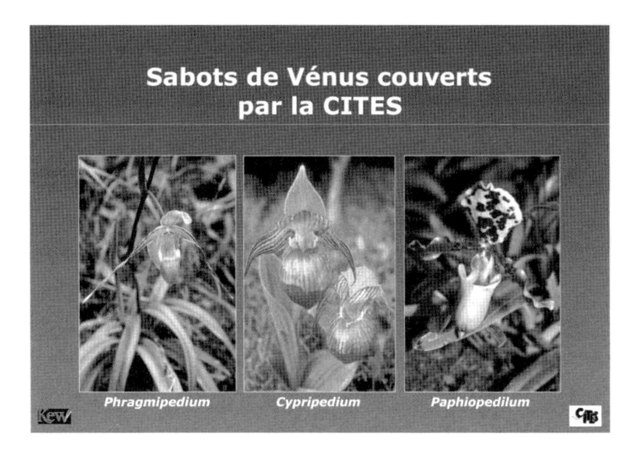

Dia 7 : Sabots de Vénus couverts par la CITES

Les genres *Paphiopedilum* et *Phragmipedium* sont inscrits à l'Annexe I de la CITES, ce qui signifie que le commerce de plantes prélevées dans la nature est interdit et que le commerce de plantes reproduites artificiellement est autorisé sous réserve de l'obtention d'un permis. Ces deux genres figurent à l'Annexe I de la CITES sous les noms de *Paphiopedilum* spp. et *Phragmipedium* spp. Cette « inscription générique » veut dire que toute espèce de ces genres nouvellement décrite est automatiquement inscrite à l'Annexe I de la CITES. Ce système garantit le contrôle immédiat de nouvelles espèces qui sont vulnérables au commerce non durable suite à leur découverte.

Le genre *Cypripedium* est inscrit à l'Annexe II de la CITES. Cela signifie que le commerce des spécimens prélevés dans la nature et de ceux reproduits artificiellement est autorisé sous réserve de l'obtention d'un permis. Néanmoins, la majorité des pays de l'aire de répartition (appelés « Etats de l'aire de répartition » dans le contexte de la CITES) ont interdit l'exportation de spécimens de *Cypripedium* prélevés dans la nature. En outre, dans leurs règlements sur le commerce des espèces sauvages qui régissent la mise en œuvre de la CITES, les 25 pays membres de l'Union européenne considèrent l'espèce *Cypripedium calceolus* comme si elle était inscrite à l'Annexe I de la Convention. De ce fait, il est rare de trouver des plantes sauvages de *Cypripedium* d'origine légale dans le commerce international. Lorsque l'on en trouve, il s'agit souvent de plantes des espèces plus communes originaires d'Amérique du Nord issues de prélèvements contrôlés ou d'opérations de sauvetage.

[Note pour l'orateur : pour consulter la liste à jour des annexes de la CITES, visitez le site Web de la CITES : www.cites.org. Vous pourrez obtenir des informations sur la réglementation de l'UE et ses mesures plus strictes à l'adresse www.eu-wildlifetrade.org. Cette diapositive montre Phragmipedium longifolium (à gauche), Cypripedium himalaicum (au milieu) et Paphiopedilum henryanum (à droite).]

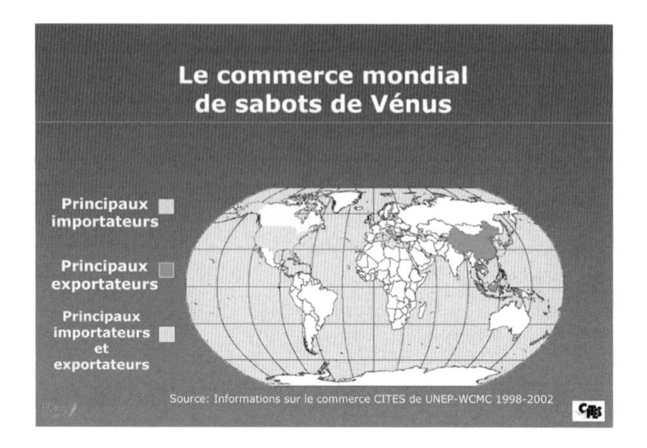

Dia 8 : Le commerce mondial de sabots de Vénus

Les sabots de Vénus, notamment le genre *Paphiopedilum*, sont l'un des cinq genres d'orchidées les plus importants en horticulture. Elles sont commercialisées en grandes quantités, principalement sous la forme de plantes vivantes de ces espèces et d'hybrides artificiels.

Les informations sur le commerce CITES indiquent que dans la période 1998–2002, plus de 660 000 sabots de Vénus ont été commercialisés à l'échelon international. Ce commerce était principalement composé de plantes reproduites artificiellement. Le commerce enregistré de plantes prélevées dans la nature, quant à lui, concernait des spécimens de *Cypripedium*.

Entre 1998 et 2002, les principaux exportateurs de sabots de Vénus reproduits artificiellement étaient Taiwan (province de Chine), l'Indonésie et la Chine. A eux seuls, ces exportateurs étaient à l'origine de plus de la moitié de toutes les exportations internationales (54%). D'autres exportateurs pendant la même période étaient la République de Corée, les Pays-Bas, la Thaïlande, les Etats-Unis, la Nouvelle Zélande, le Japon et la Belgique. Chacun d'entre eux a exporté plus de 10 000 plantes.

Le plus grand importateur de sabots de Vénus reproduits artificiellement entre 1998 et 2002 était le Japon, qui à lui seul représente plus de la moitié des importations globales (56%). Pendant la même période, d'autres importateurs étaient le Canada, la République de Corée, les Etats-Unis, l'Allemagne, l'Italie et la Suisse. Chacun de ces pays a importé plus de 10 000 plantes.

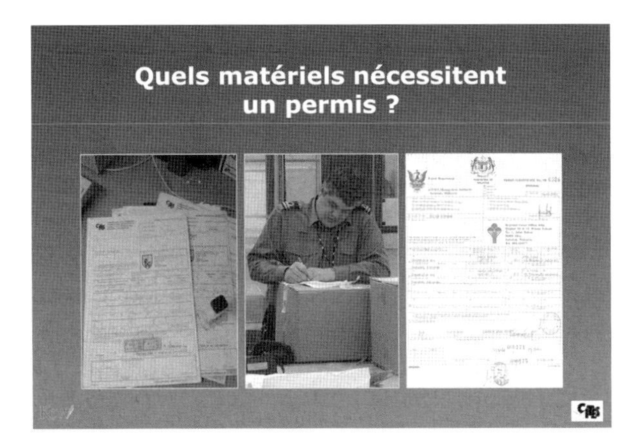

Dia 9 : Quels matériels requièrent un permis ?

La réponse la plus simple est de dire que tout matériel requiert un permis à moins qu'il fasse l'objet d'une dérogation.

Les contrôles CITES s'appliquent aux plantes « vivantes ou mortes » et à « toute partie ou tout produit facilement identifiable ».

Cela signifie que ce ne sont pas seulement les plantes elles-mêmes qui sont contrôlées mais aussi leurs parties – graines, boutures et feuilles. Les produits obtenus à partir de la plante peuvent eux aussi être contrôlés, et si le nom d'une espèce inscrite aux annexes de la CITES est écrit sur l'emballage, le produit est censé le contenir et peut donc être contrôlé.

Dans le cas des espèces inscrites à l'Annexe I de la CITES, toute la plante ainsi que ses parties et produits sont contrôlés – qu'ils soient vivants ou morts. Toutefois, il existe une dérogation : les cultures de plantules ou de tissus « obtenues *in vitro* en milieu solide ou liquide et transportées en conteneurs stériles » sont exclues des contrôles. Pour bénéficier de cette dérogation, il n'est pas nécessaire que le matériel se trouve dans des flacons ou des bouteilles conventionnels, mais tout simplement dans des conteneurs stériles. Cependant, pour que le matériel puisse faire l'objet de la dérogation, l'origine de celui-ci doit être licite. Nous verrons les contrôles relatifs aux cultures de tissus dans plus de détails dans la prochaine diapositive.

Quant aux espèces inscrites à l'Annexe II de la CITES, les plantes « vivantes ou mortes » sont contrôlées, ainsi que toute partie ou tout produit facilement identifiable précisé dans les annexes. Dans le cas de *Cypripedium*, les seuls parties et produits exclus des contrôles de la CITES sont : a) les graines et le pollen (y compris les pollinies) ; b) les cultures de plantules ou de tissus obtenues *in vitro* en milieu solide ou liquide et transportées en conteneurs stériles ; ainsi que c) les fleurs coupées de plantes reproduites artificiellement.

[Note pour l'orateur : la diapositive montre des permis CITES et un douanier en train de vérifier des documents.]

Dia 10

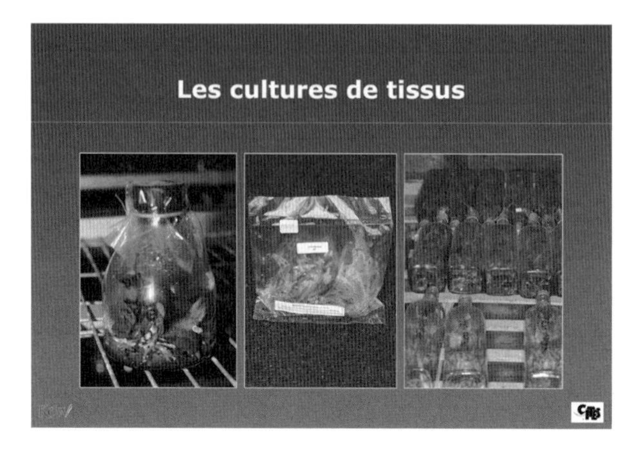

Dia 10 : Les cultures de tissus

Suite à l'inscription des genres *Paphiopedilum* et *Phragmipedium* à l'Annexe I, la Conférence des Parties (CdP) à la CITES a décidé d'exclure des contrôles les cultures de tissus de ces genres. Il s'agit d'une dérogation unique, c'est à dire, l'on exclut des contrôles « une partie ou un produit facilement identifiable » d'un spécimen inscrit à l'Annexe I. Plusieurs Parties se sont opposées à la dérogation, qui a cependant été adoptée à la majorité au moyen d'un vote. Le but de cette dérogation était d'encourager la reproduction artificielle de ces plantes très recherchées afin de réduire la pression sur les populations sauvages.

A l'époque, on croyait que la reproduction artificielle ne pourrait jamais nuire aux populations sauvages, mais il s'est avéré récemment que ce n'est pas le cas. Tout particulièrement, le prélèvement illicite et non durable d'espèces endémiques du Viêt Nam a éveillé une préoccupation considérable. Plusieurs pays ont manifesté leur inquiétude que la dérogation pour les plantules en flacons et les cultures de tissus ait été utilisée par des commerçants sans scrupules pour « légaliser » le commerce de plantes dont la population parentale avait été illégalement prélevée dans la nature.

La population parentale utilisée dans la production des cultures de tissus doit avoir été obtenue légalement d'après la réglementation sur l'espèce en question dans son pays d'origine. Si cette population parentale est illicite, aucune culture de tissus qui en découle ne fera l'objet de la dérogation des contrôles de la CITES. Par conséquent, ce matériel pourra être confisqué par les autorités chargées de la lutte contre la fraude.

[Note pour l'orateur : cette diapositive montre des cultures de plantules stériles dans des flacons, des bouteilles et des sachets.]

Sabots de Vénus
couverts par la CITES

Dia 12

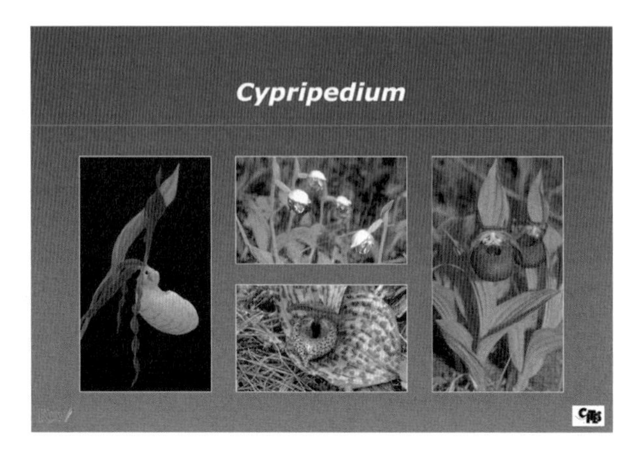

Dia 12 : ***Cypripedium***

Le genre *Cypripedium* comprend quelque 50 espèces qui poussent dans les régions tempérées de l'hémisphère Nord en Asie, en Europe et en Amérique du Nord, et dont les latitudes les plus méridionales de la répartition sont le Honduras, le Guatemala et le sud de la Chine. Ces espèces poussent dans des habitats très variés tels que des forêts de conifères, des forêts mixtes avec des feuillus, des marais ou des prairies, entre autres. Ce sont des plantes terrestres avec des feuilles qui, chez la plupart des espèces, poussent à nouveau depuis la base tous les ans. Les fleurs ont la forme d'un sabot et présentent des couleurs très diverses telles que le vert, le blanc, le jaune, le rouge ou le pourpre intense. Les *Cypripedium* sont attrayantes pour les amateurs de plantes des climats tempérés parce qu'elles présentent toutes un certain degré de résistance au froid et peuvent pousser à l'extérieur pendant au moins une partie de la saison.

[Note pour l'orateur : cette diapositive montre Cypripedium parviflorum (à gauche), Cypripedium guttatum (en haut au milieu), Cypripedium lichiangense (en bas au milieu) et Cypripedium smithii (à droite).]

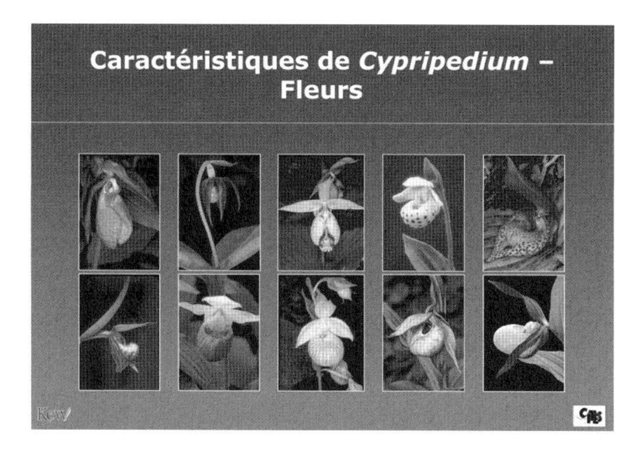

Dia 13 : Caractéristiques de *Cypripedium* – Fleurs

Toutes les fleurs de *Cypripedium* ont la forme d'un sabot. Il s'agit de fleurs individuelles ou de groupes de fleurs, dont les couleurs varient entre le vert, le pourpre avec des taches blanches, le jaune doré et le marron pourpre.

Les fleurs ne tombent pas mais persistent sur le fruit.

[Note pour l'orateur : La diapositive montre Cypripedium acaule (en haut à gauche), Cypripedium palangshanense (en haut, la deuxième image en commençant par la gauche), Cypripedium japonicum (en haut au milieu), Cypripedium wardii (en haut, la deuxième image en commençant par la droite), Cypripedium lichiangense (en haut à droite), Cypripedium arietinum (en bas à gauche), Cypripedium reginae (en bas, la deuxième image en commençant par la gauche), Cypripedium irapeanum (en bas au milieu), Cypripedium x froschii (en bas, la deuxième image en commençant par la droite) et Cypripedium calceolus (en bas à droite).]

Dia 14

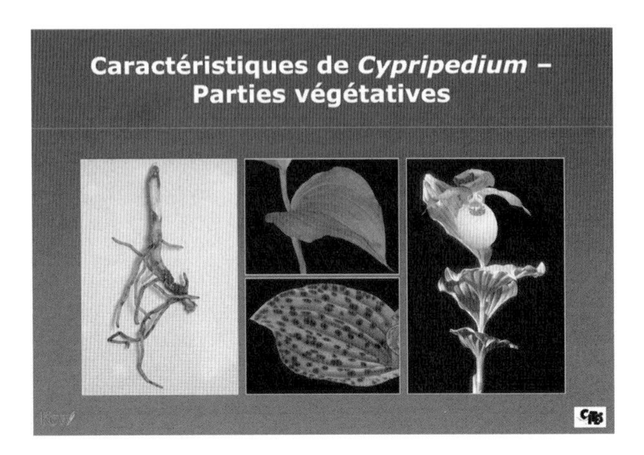

Dia 14 : Caractéristiques de *Cypripedium* – Parties végétatives

- Toutes les plantes du genre *Cypripedium* possèdent une tige souterraine caractéristique qui ressemble à une racine et qui porte le nom de rhizome.
- Chez la plupart des espèces, ce rhizome est court et rarement ramifié et forme une chaîne de marques de croissance annuelles qui ressemble à un chapelet.
- La plante perd ses feuilles tous les ans, et il ne reste que le rhizome pendant la période de repos. Au printemps la plante pousse de nouveau à partir des bourgeons du rhizome.
- Les feuilles sont généralement ovales.
- Les feuilles sont plissées depuis la base jusqu'à leur extrémité et présentent des nervures proéminentes.
- Les feuilles ont souvent des poils, notamment sur les nervures et les bords.

[Note pour l'orateur : la diapositive montre un rhizome avec un bourgeon et des racines (à gauche), des feuilles aux couleurs variées (en haut et en bas au milieu), des feuilles sur une tige et la fleur de Cypripedium fasciolatum (à droite).]

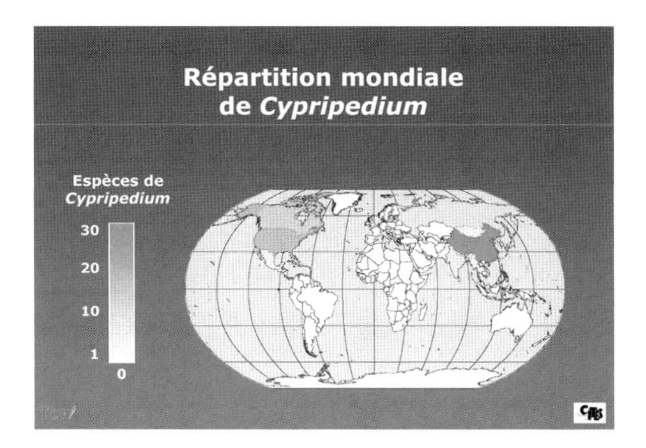

Dia 15 : Répartition mondiale de *Cypripedium*

Quoique le genre *Cypripedium* soit réparti dans les régions tempérées de l'hémisphère Nord, la Chine est de loin le pays qui contient le plus grand nombre d'espèces. En raison du manque d'accès à ces plantes dans le passé, ces espèces sont à présent extrêmement recherchées. La plupart des espèces que l'on cultive couramment sont celles originaires d'Amérique du Nord, l'espèce européenne *Cypripedium calceolus* et l'espèce japonaise *Cypripedium formosanum.*

Dia 16

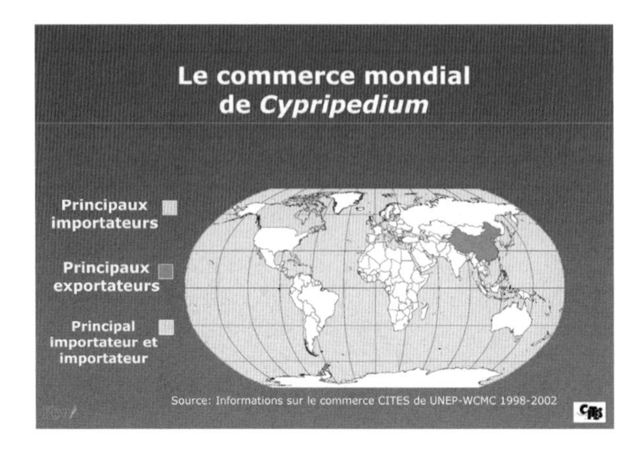

Dia 16 : Le commerce mondial de *Cypripedium*

La grande majorité du commerce de plantes du genre *Cypripedium* enregistré par la CITES pendant la période 1998–2002 concernait des plantes vivantes. D'après les rapports, presque toutes les plantes commercialisées (>98 %) avaient été reproduites artificiellement.

Les rapports indiquent que les principaux fournisseurs de matériel reproduit artificiellement étaient Taiwan (province de Chine), la Chine et la République de Corée. Ces trois exportateurs ont fourni 93 pour cent du matériel déclaré comme reproduit artificiellement entre 1998 et 2002.

Pendant cette période, le principal importateur de *Cypripedium* était le Japon, qui a reçu 78 pour cent des importations enregistrées par la CITES. Le Japon, la République de Corée, le Canada et l'Allemagne ont importé 95 pour cent de tout le matériel commercialisé.

Un peu moins de 2 pour cent du commerce enregistré entre 1998 et 2002 concernait des plantes prélevées dans la nature. Les Etats-Unis, la Fédération Russe et la Chine étaient les principaux fournisseurs de ce matériel, tandis que l'Allemagne, le Japon et le Royaume-Uni en étaient les principaux importateurs.

[Note pour l'orateur : les informations sur le commerce CITES peuvent être téléchargées à partir de la base de données sur le commerce CITES de UNEP-WCMC (CITES Trade Database, en anglais). Vous pourrez y accéder en ligne à travers le site Web du Secrétariat CITES : www.cites.org]

Dia 17 : *Cypripedium lichiangense* et *C. palangshanense*

L'espèce *Cypripedium lichiangense* a été décrite pour la première fois en 1994 dans le nord-est du Myanmar et dans le nord-ouest de Yunnan et le sud-ouest de Sichuan en Chine. Cette espèce, très convoitée, possède deux feuilles qui abritent une fleur tâchetée comme les œufs de certains oiseaux. Il est particulièrement difficile de cultiver et de reproduire cette espèce, et les plantes cultivées survivent rarement plus de 3-4 ans.

Quoique la description de l'espèce *Cypripedium palangshanense* date de 1936, cette espèce est demeurée pratiquement inconnue pour le monde occidental jusqu'en 1998, lorsqu'elle a été redécouverte. Il s'agit d'un espèce endémique d'une zone très restreinte du nord-ouest et de l'est de Sichuan en Chine. Cette plante possède un rhizome très fin et rampant et des petites fleurs pourpres.

Ces deux espèces sont désormais touchées par le commerce.

[Note pour l'orateur : la diapositive montre Cypripedium lichiangense (à gauche) et Cypripedium palangshanense (en haut et en bas à droite).]

Dia 18

Dia 18 : *Phragmipedium*

Le genre *Phragmipedium* ne comprend que quelque 20 espèces, réparties du sud du Mexique et du Guatemala en passant par l'Amérique centrale jusqu'au sud de la Bolivie et du Brésil en Amérique du Sud. Certaines de ces espèces poussent au sol, alors que d'autres poussent sur des rochers ou des arbres, qu'elles utilisent comme support. On les trouve généralement autour de chutes d'eau et dans d'autres zones humides.

Ces plantes possèdent un rhizome – une tige modifiée qui ressemble à une racine – rampant et des feuilles coriaces en forme de V en coupe transversale. Elles sont moins populaires dans le commerce que leurs rivales du genre *Paphiopedilum*, issues d'Asie du Sud-Est. Toutefois, la découverte de l'espèce *Phragmipedium besseae* avec ses insolites fleurs rouges a fortement éveillé dans les années 1980 l'interêt commercial et celui des amateurs pour ce genre. Plus récemment, en 2002, la découverte au Pérou de *Phragmipedium kovachii*, aux grandes fleurs mauves, a stimulé encore plus l'intérêt pour ce groupe.

[Note pour l'orateur : la diapositive montre Phragmipedium besseae (à gauche et en haut au milieu) et Phragmipedium longifolium (à droite et en bas au milieu)]

Dia 19 : Caractéristiques de *Phragmipedium* – Fleurs

Lorsque les fleurs se trouvent au stade de boutons floraux, les bords des sépales – les feuilles modifiées qui ressemblent à des pétales – qui entourent la fleur se touchent, tandis que ceux-ci se chevauchent chez les autres groupes. En outre, le bord du labelle de la fleur en forme de sabot est replié vers l'intérieur. Contrairement aux *Cypripedium*, lorsque les fleurs se flétrissent, elles se détachent de l'extrémité des fruits.

[Note pour l'orateur : la diapositive montre Phragmipedium besseae var. dalessandroi (à gauche), Phragmipedium kovachii (à gauche au milieu, en haut), Phragmipedium longifolium (à droite au milieu, en haut), Phragmipedium lindenii (à droite), Phragmipedium schlimii (à gauche au milieu, en bas)et Phragmipedium wallisii (à droite au milieu, en bas).]

Dia 20

Dia 20 : Caractéristiques de *Phragmipedium* – Parties végétatives

Les feuilles de *Phragmipedium* sont oblongues ou linéaires et généralement vert mat ou brillant sans marques. Elles sont plates et coriaces et possèdent une nervure médiane proéminente mais pas de nervures secondaires évidentes. Elles sont disposées en deux rangées de chaque côté de la tige. Au lieu de perdre ses feuilles une fois par an, la plante garde ses feuilles pendant au moins deux ans. Les feuilles sont habituellement longues, en forme de ruban et terminées en pointe, tandis que les tiges sont généralement courtes. Toutes les espèces possèdent un rhizome court ou rampant, dépourvu des « anneaux » de croissance annuels que l'on trouve chez les *Cypripedium*.

[Note pour l'orateur : la diapositive montre Phragmipedium besseae (à gauche), Phragmipedium lindleyanum (en haut au milieu), Phragmipedium longifolium (en bas au milieu) et les feuilles terminées en pointe d'une espèce de Phragmipedium (à droite).]

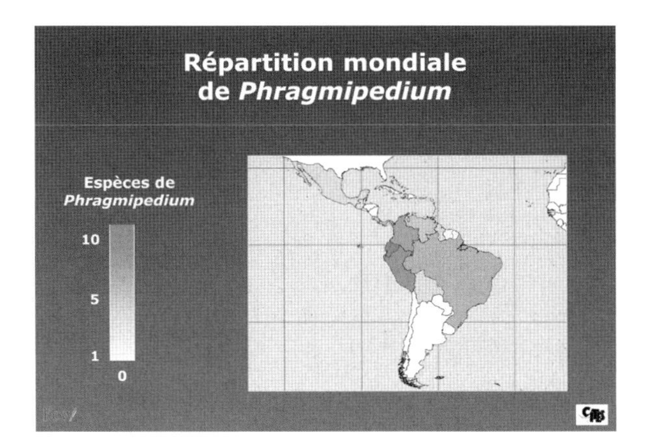

Dia 21 : Répartition mondiale de *Phragmipedium*

Ce genre est réparti en Amérique centrale et en Amérique du Sud, du sud du Mexique et du Guatemala jusqu'au sud de la Bolivie et du Brésil.

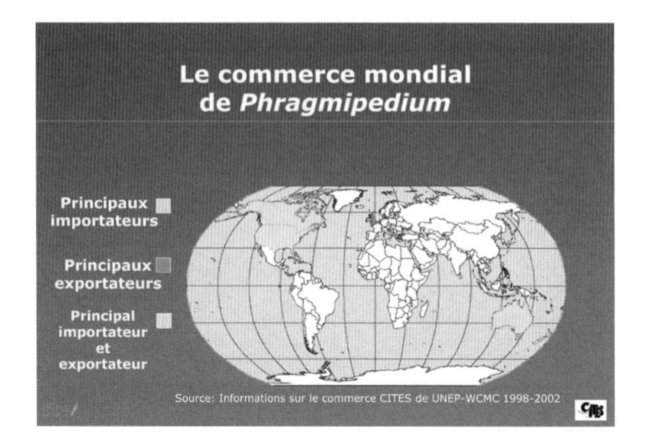

Dia 22 : Le commerce mondial de *Phragmipedium*

Le commerce de taxons de *Phragmipedium* enregistré par la CITES pendant la période 1998–2002 représente moins de 2 pour cent de tout le commerce international de sabots de Vénus. Ce commerce ne concernait que des spécimens reproduits artificiellement.

Pendant la même période, l'Equateur, le Royaume-Uni, Taiwan (province de Chine) et les Etats-Unis étaient les principaux exportateurs de matériel reproduit artificiellement, ce qui représente 84 pour cent du total. Le Japon, les Etats-Unis, le Canada et l'Australie totalisaient 77 pour cent de tout le matériel de *Phragmipedium* importé dans cet espace de temps. Les informations sur le commerce CITES pour la période en question indiquent également que bon nombre de pays ont importé des plantes réexportées par les Etats-Unis.

[Note pour l'orateur : les informations sur le commerce CITES peuvent être téléchargées à partir de la base de données sur le commerce CITES de UNEP-WCMC (CITES Trade Database, en anglais). Vous pourrez y accéder en ligne à travers le site Web du Secrétariat CITES : www.cites.org]

Dia 23 : *Phragmipedium kovachii*

La découverte et l'importation aux Etats-Unis de l'éspèce *Phragmipedium kovachii* ont généré une forte publicité. L'espèce a été découverte dans le nord du Pérou et décrite en 2002. Dans leur intérêt pour cette nouvelle espèce, les média ont oublié de souligner que l'espèce aurait été prélevée à outrance et menée à l'extinction par les collectionneurs. Cependant, au moins un des cinq sites connus existait encore en 2004 et contenait plusieurs centaines de plantes. D'après certaines sources, le gouvernement péruvien aurait autorisé un prélèvement très limité dans la nature afin de permettre la reproduction contrôlée au moyen de graines. Le gouvernement permettrait l'exportation de certaines lignées de l'espèce et d'hybrides issus au moins d'une pépinière approuvée. La conservation de cette espèce insolite et recherchée ne pourrait que bénéficier de la reproduction à grande échelle de plantules le plus rapidement possible, afin de réduire la valeur des plantes illicites prélevées dans la nature.

[Note pour l'orateur : cette diapositive montre trois images de Phragmipedium kovachii.]

Dia 24

Dia 24 : *Paphiopedilum*

Paphiopedilum est le genre le plus vaste de sabots de Vénus, car il contient quelque 80 espèces réparties dans les régions tropicales d'Asie, du sud de l'Inde jusqu'en Nouvelle Guinée et dans les Philippines. On peut les trouver poussant au sol, dans des endroits rocailleux ou bien accrochées à des arbres ou à d'autres plantes. La plupart des espèces poussent au sol, dans des feuilles en décomposition ou dans des fissures de rochers qui contiennent de la matière organique. Elles sont présentes dans des habitats très variés qui vont des branches de grands arbres dans les forêts tropicales de la Thaïlande jusqu'aux durs sols de serpentine du mont Kinabalu en Indonésie.

Toutes les espèces de ce genre présentent une fleur caractéristique en forme de sabot, et dans le cas de certaines espèces les fleurs sont exagérément enflées, comme l'orchidée « Bubble gum », *Paphiopedilum micranthum,* à droite sur la diapositive. Il s'agit du groupe de sabots de Vénus le plus apprécié par les collectionneurs et les horticulteurs. En outre, *Paphiopedilum* est l'un des cinq genres les plus importants d'orchidées en horticulture et la base de bon nombre d'hybrides artificiels que l'on obtient et produit chaque année. Cependant, certaines espèces du genre *Paphiopedilum* ont fait l'objet d'un niveau considérable de prélèvements et de commerce illicites.

[Note pour l'orateur : cette diapositive montre Paphiopedilum exul (à gauche), Paphiopedilum rothschildianum (au milieu) et Paphiopedilum micranthum (à droite).]

Dia 25 : Caractéristiques de *Paphiopedilum* – Fleurs

Lorsque les fleurs se trouvent au stade de boutons floraux, les bords des sépales – les feuilles modifiées qui ressemblent à des pétales – qui entourent la fleur se chevauchent. En outre, le bord du labelle de la fleur en forme de sabot n'est pas replié vers l'intérieur. Contrairement aux *Cypripedium*, lorsque les fleurs se flétrissent, elles se détachent de l'extrémité des fruits. Dans le cas de bon nombre d'espèces, les fleurs ont des poils et des protubérances qui ressemblent à des verrues. De nombreuses espèces ne présentent qu'une seule fleur, mais certaines en produisent plusieurs. La couleur des fleurs varie considérablement du vert au blanc, de l'or au pourpre. Certaines des espèces les plus appréciéees produisent des fleurs avec des pétales longs et pendants qui, dans le cas de *Paphiopedilum sanderianum*, peuvent dépasser un mètre de longueur.

[Note pour l'orateur : la diapositive montre Paphiopedilum malipoense (à gauche), Paphiopedilum haynaldianum (à gauche au milieu, en haut), Paphiopedilum bellatulum (à droite au milieu, en haut), Paphiopedilum druryi (à gauche au milieu, en bas), Paphiopedilum liemianum (à droite au milieu, en bas) et Paphiopedilum sanderianum (à droite).]

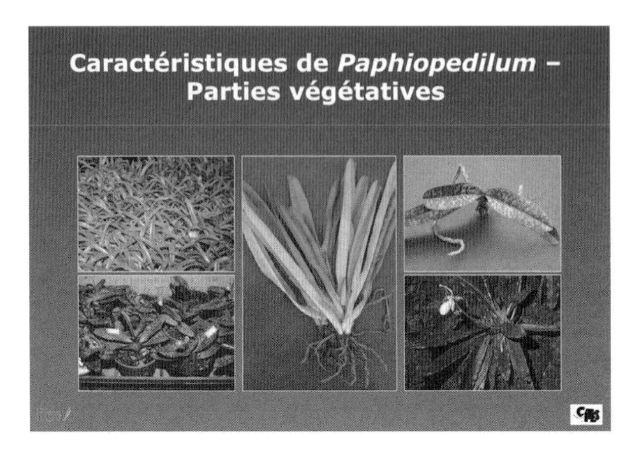

Dia 26 : Caractéristiques de *Paphiopedilum* – Parties végétatives

Toutes les espèces possèdent un rhizome, qui est généralement court. Un nombre réduit d'espèces, *P. bullenianum, P. armeniacum, P. micranthum* et *P. druryi* possèdent des rhizomes rampants qui peuvent atteindre un mètre de longueur.

Les feuilles de *Paphiopedilum* sont coriaces, avec une nervure médiane proéminente. Les feuilles ont la forme d'un V en coupe transversale, et peuvent être courtes en forme de ruban, oblongues ou linéaires. En général, les feuilles sont courtes, de moins de 20 centimètres de longueur. Il existe cependant une exception à cette règle : le groupe multifleurs, qui contient des espèces telles que *Paphiopedilum sanderianum, P. rothschildianum* et *P. lowii*. On classe la couleur des feuilles en différents types, ce qui peut être très utile pour l'identification des plantes. La couleur oscille entre vert mat ou brillant et pourpre marbré.

[Note pour l'orateur : la diapositive montre des plantes avec des feuilles vertes coriaces et en forme de ruban (en haut à gauche), des feuilles marbrées (en bas à gauche), des feuilles de plus de 20 cm de longueur (au milieu), des feuilles marbrées (en haut à droite) et la plante dans son habitat avec des feuilles en forme de ruban vert brillant (Paphiopedilum liemianum, en bas à droite).]

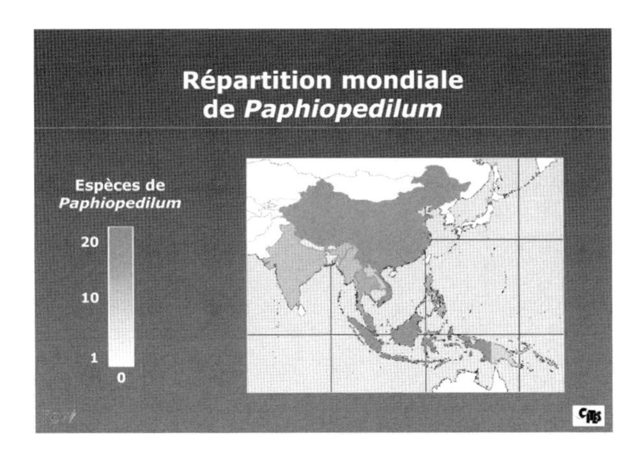

Dia 27 : Répartition mondiale de *Paphiopedilum*

Le genre *Paphiopedilum* comprend quelque 80 espèces. Alors que bon nombre d'entre elles sont des espèces endémiques de zones très restreintes, le genre dans son ensemble est réparti en Asie du Sud-Est, d'Inde jusqu'aux Philippines en passant par la Chine et jusqu'en Nouvelle Guinée et les îles Salomon en passant par l'archipel malais. A l'heure actuelle, bon nombre des espèces les plus recherchées se trouvent en Chine et au Viêt Nam.

Dia 28

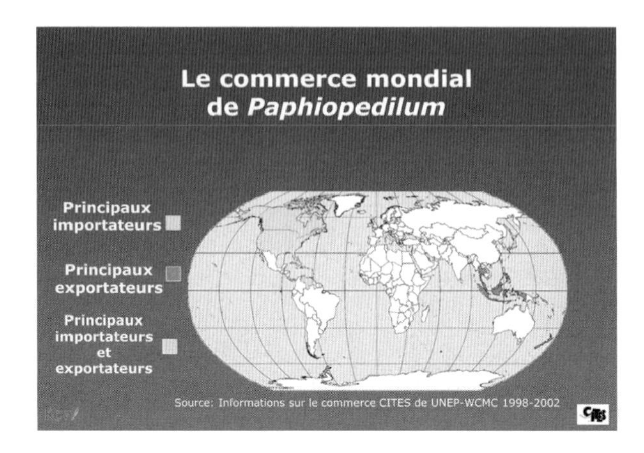

Dia 28 : Le commerce mondial de *Paphiopedilum*

Entre 1998 et 2002, de nombreux pays ont déclaré des exportations de taxons du genre *Paphiopedilum* reproduits artificiellement. L'Indonésie, les Pays-Bas, la Thaïlande et la Nouvelle-Zélande étaient les principaux pays exportateurs, dont les transactions représentaient 79 pour cent de toutes les exportations enregistrées. De plus, Taiwan (province de Chine), les Etats-Unis, le Japon et la Belgique ont exporté plus de 10 000 plantes chacun. Les 8 exportateurs principaux étaient responsables de 90 pour cent de toutes les exportations. Si l'on ajoute le Japon, la Malaisie et l'Autriche, ce chiffre dépasse 97 pour cent de toutes les exportations.

Pendant la même période, les principaux exportateurs étaient le Japon et les Etats-Unis, dont les transactions représentaient 75 pour cent de toutes les importations documentées. Si l'on ajoute le Canada, l'Italie, la Suisse, Hong Kong et le Venezuela, ces pays ont été responsables de 93 pour cent des importations.

[Note pour l'orateur : les informations sur le commerce CITES peuvent être téléchargées à partir de la base de données sur le commerce CITES de UNEP-WCMC (CITES Trade Database, en anglais). Vous pourrez y accéder en ligne à travers le site Web du Secrétariat CITES : www.cites.org]

Dia 29 : La Chine et le Viêt Nam

Pendant les années 1980 et au début des années 1990, un nouveau groupe de *Paphiopedilum* a été découvert en Chine. Non seulement il s'agissait de nouvelles espèces intéressantes et différentes de toutes les autres que l'on connaissait déjà, mais elles représentaient également toute une lignée potentielle pour la création d'hybrides. *Paphiopedilum armeniacum, P. emersonii, P. micranthum* et *P. malipoense* sont certaines de ces nouvelles espèces. Cette découverte a donné lieu à un niveau considérable de commerce illicite. Cependant, ces espèces sont à présent bien établies en culture, ce qui a réduit le besoin de prélever des plantes dans la nature, mais les nouvelles formes de couleur telles que les albinos sont toujours recherchées. Toutefois, la pression des prélèvements destinés au commerce international a rendu ces espèce vulnérables à l'extinction en Chine.

A la fin des années 1990 et au début des années 2000, plusieurs espèces nouvelles et insolites de sabots de Vénus ont été décrites au Viêt Nam, notamment *Paphiopedilum vietnamense* et *P. hangianum*. Ces deux espèces sont très recherchées en raison de leurs formes nouvelles. L'espèce *Paphiopedilum vietnamense* a été décrite en 1999. En 2001, lors d'une expédition organisée pour réaliser une étude de terrain sur le seul site connu, on n'y a trouvé que quelques plantules. Cette espèce est désormais considérée En danger critique d'extinction en raison de sa répartition très restreinte et de son niveau d'exploitation.

[Note pour l'orateur : la diapositive montre Paphiopedilum armeniacum (à gauche), Paphiopedilum emersonii (en haut au milieu), Paphiopedilum malipoense (en bas au milieu) et Paphiopedilum micranthum (à droite).]

Dia 30 : Sabots de Vénus couverts par la CITES : Résumé

Dans cette section, nous avons passé en revue:

- les trois genres de sabots de Vénus couverts par la CITES : *Cypripedium, Phragmipedium* et *Paphiopedilum,* ainsi que

- leurs caractéristiques, répartition et commerce mondial.

[Note pour l'orateur : la diapositive montre Phragmipedium wallisii (à gauche), Cypripedium flavum (au milieu) et Paphiopedilum helenae (à droite).]

Mise en œuvre de la
CITES pour les sabots
de Vénus

Dia 32

Dia 32 : Mise en œuvre de la CITES et lutte contre la fraude

Les contrôles de la CITES se mettent en œuvre à différents niveaux. Dans les pays exportateurs, ils impliquent l'inspection de pépinières, de commerçants, de marchés et moins souvent – quoique ce soit le point le plus important – l'inspection des plantes au moment de l'exportation. Des inspections peuvent également avoir lieu au moment de l'importation et post-importation dans les pays avec de grands volumes de commerce. Les organes chargés du respect de la loi surveillent également les expositions commerciales et la publicité dans la presse spécialisée et sur Internet.

Peu sont les pays où le personnel chargé de l'application de la loi possède une formation spécialisée pour pouvoir identifier des spécimens CITES, que ce soit des plantes ou des animaux. En général, le respect de la Convention est assuré, dans les cas des plantes, par des douaniers non spécialisés ou par des fonctionnaires avec une formation orientée vers les contrôles phytosanitaires. Lorsque les contrôles sont menés par des douaniers « généralistes », la démarche se concentre sur la documentation plutôt que sur les plantes. Ainsi, les douaniers peuvent vérifier si les permis sont remplis correctement, approuvés, s'ils portent un sceau et s'ils ont été délivrés par les autorités pertinentes. Ils examinent également d'autres documents et des factures afin de vérifier si d'autres matériels CITES mentionnés dans les documents connexes ne figurent pas sur les permis CITES.

Lorsque ce sont des douaniers avec une formation générale qui doivent contrôler les plantes soumises aux dispositions de la CITES, il est essentiel que ces personnes disposent d'un centre spécialisé dans l'identification et la conservation des plantes. L'idéal serait qu'il s'agisse de l'autorité scientifique CITES. Cependant, l'autorité scientifique est parfois un comité ou un département du gouvernement dont le savoir-faire se concentre sur les espèces animales. Dans ce cas, les autorités chargées de l'application de la loi devraient établir de bons rapports avec un jardin botanique ou un herbier local ou national. De tels contacts sont essentiels.

Les douaniers auront besoin de formation de base sur les plantes et les parties et produits contrôlés par la CITES, ainsi que d'aide pour déceler le commerce préjudiciable. Surtout, les douaniers doivent avoir accès à des experts qui puissent identifier les plantes CITES. De tels experts peuvent également fournir des conseils sur le matériel confisqué et avoir accès à des installations pour garder les plantes saisies ou confisquées. Ces scientifiques peuvent être convoqués comme témoins experts, essentiels si l'infraction donne lieu à des poursuites judiciaires et à un procès.

[Note pour l'orateur : pour savoir quel est le membre du personnel du Secrétariat CITES que vous devriez contacter sur des questions relevant de la mise en œuvre de la CITES et la lutte contre la fraude, veuillez consulter la liste du personnel sur le site Web de la CITES : www.cites.org.]

Dia 33

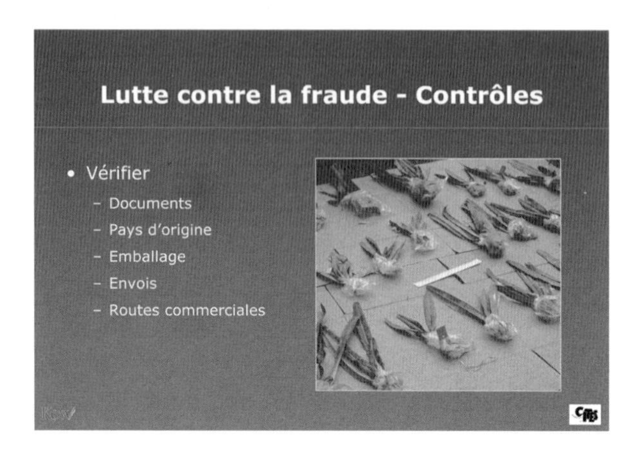

Dia 33 : Lutte contre la fraude - Contrôles

Documents – Vérifier l'authenticité des permis CITES (signatures, sceaux) et comparer les noms et le nombre de spécimens figurant sur le permis avec le bulletin de livraison ou la facture. Vérifier également la source des plantes – sont-elles déclarées comme étant prélevées dans la nature ou reproduites artificiellement ? Utiliser les bases de données et les listes de référence recommandées dans la section des références et ressources. S'agit-il de plantules en flacons ou de cultures de tissus censées être exclues des contrôles de la CITES ? Le cas échéant, et s'il s'agit d'espèces récemment décrites, il convient de demander à votre organe de gestion de vérifier si l'origine de la population parentale est licite.

Pays d'origine – Vérifiez toujours le pays d'origine qui figure sur le permis. Les plantes sont-elles exportées par un pays de l'aire de répartition de ces espèces ? Si c'est le cas, les plantes ont plus de chances d'avoir été prélevées dans la nature. Les pays peuvent exprimer leur préoccupation quant à l'exportation illégale de leurs sabots de Vénus prélevés dans la nature et demander l'assistance d'autres pays Partie et non Partie à la CITES pour contrôler ce commerce. Ces demandes sont généralement publiées sous la forme d'une Notification aux Parties à la CITES (vous trouverez ces informations sur le site Web de la CITES : www.cites.org). Le Viêt Nam, par exemple, est un pays qui a exprimé sa préoccupation quant au commerce international des espèces indigènes de *Paphiopedilum.*

Emballage – En général, les pépinières emballent et conditionnent les plantes avec soin afin d'éviter qu'elles soient endommagées. Ensuite, les plantes sont expédiées dans des caisses qui portent le nom de l'établissement et des étiquettes imprimées. Les envois de plantes prélevées illégalement peuvent être mal emballés avec des matériaux locaux, et contenir des étiquettes manuscrites (parfois avec des informations sur le prélèvement) qui ne sont identifiées qu'au niveau générique, pour cacher le fait que de nouvelles espèces non mentionnées puissent avoir été prélevées.

Envois de plantes – Les collections de plantes illicites se composent généralement de petits échantillons de plantes de tailles et d'ages différents dont la forme n'est pas homogène. Parfois, les plantes sont endommagées (leurs racines sont cassées), et les tiges et les racines contiennent de la terre, des mauvaises herbes ou des plantes indigènes. Les plantes reproduites artificiellement sont de taille et de forme homogène, et sont exemptes de terre, de parasites et de maladies, de mauvaise herbes et de plantes indigènes.

Route commerciales et contrebande – Parfois, des collections illicites d'espèces rares ou nouvelles sont expédiées à travers la poste ou les services de messagerie ou même dans les bagages à main de passagers afin de passer inaperçues. Les collections son parfois divisées en plusieurs lots et expédiés dans plusieurs colis pour augmenter leurs chances de survie et la probabilité qu'au moins certaines de ces plantes échapperont aux contrôles.

Dia 34

Dia 34 : Comment distinguer les plantes prélevées dans la nature de celles reproduites artificiellement ?

Ce n'est pas facile. Cependant, certaines caractéristiques sont utiles pour faire cette distinction.

Les plantes prélevées dans la nature portent les marques distinctives d'avoir poussé dans leur habitat naturel. Par contre, les plantes cultivées dans les pépinières portent les marques d'un environnement artificiel et bien contrôlé. Elles sont propres, uniformes et très bien conditionnées dans leur emballage. Parfois, les orchidées sont cultivées en plein air ou dans des installations simples visant à les protéger de l'excès de soleil, et peuvent donc avoir certaines traces semblables à celles des plantes prélevées dans la nature. Lorsque vous soupçonnez qu'une plante peut avoir été prélevée dans la nature plutôt que reproduite artificiellement, il est donc important de demander à un expert de le vérifier.

Le Volume I du *Manuel d'identification CITES sur la flore*, disponible auprès du Secrétariat CITES, contient des informations détaillées qui permettent de distinguer les plantes prélevées dans la nature de celles reproduites artificiellement pour les principaux groupes de plantes couverts par la Convention. Cependant, n'oubliez pas de toujours demander à un expert de vérifier si vous avez raison !

Dia 35 : Les orchidées prélevées dans la nature

Les racines des plantes prélevées dans la nature sont souvent mortes, grossièrement cassées ou coupées afin de nettoyer la plante après le prélèvement. Parfois, de nouvelles racines poussent à partir de vieilles racines endommagées. On peut également trouver des restes du substrat naturel attaché aux racines des plantes prélevées dans la nature. Observez également si le substrat attaché aux racines est disposé en suivant un certain ordre. Par exemple, les racines peuvent être directement en contact avec de la matière organique, entourée de tourbe de sphaigne utilisée pour le transport et enfin de compost utilisé en horticulture tel que de l'écorce ou de la laine de roche. Toutefois, n'oubliez pas d'être toujours prudent dans votre évaluation.

Les feuilles des plantes prélevées dans la nature conservent les traces de leur habitat naturel et des dommages causés par le prélèvement. Souvent, on voit également un contraste entre ces parties et la nouvelle croissance qui a eu lieu après le prélèvement. Les feuilles basales sont souvent mortes ou endommagées. Les feuilles peuvent avoir une surface piquetée à cause de la dessiccation, et présenter les marques d'insectes. Sur les plantes récemment prélevées, il peut également y avoir de la mousse, des lichens ou des hépatiques. Ce genre d'organismes ne survit pas en général dans les conditions contrôlées d'une pépinière d'orchidées. Au fur et à mesure que les plantes poussent après leur arrivée dans une pépinière, de nouvelles feuilles commencent à pousser. Celles-ci sont propres et nouvelles, et contrastent fortement avec les vieilles feuilles « sauvages ». Toutefois, ces vieilles feuilles visiblement « sauvages » peuvent avoir été enlevées intentionnellement pour ne laisser que les nouvelles feuilles qui ont poussé après l'arrivée de la plante dans la pépinière.

Le Volume I du *Manuel d'identification CITES sur la flore* contient des informations détaillées sur la manière de distinguer les plantes prélevées dans la nature de celles reproduites artificiellement. Cependant, il est toujours important de demander l'avis d'un expert pour confirmer que les plantes que vous avez identifié comme étant prélevées dans la nature le sont vraiment. Les plantes cultivées dans de mauvaises conditions en plein air ou dans des installations visant à les protéger de l'excès de soleil présentent parfois certaines des traces que l'on trouve sur les plantes prélevées dans la nature.

Dia 35

[Note pour l'orateur : le Comité pour les plantes a élaboré une série de directoires régionaux qui contiennent les coordonnées d'experts en matière de CITES dans les différents pays (visitez le site Web de la CITES pour plus de détails). Vous pourrez utiliser ces informations pour contacter un expert en la matière. Le Volume 1 du Manuel d'identification CITES sur la flore contient des informations sur les caractéristiques des orchidées prélevées dans la nature et de celles reproduites artificiellement. Le Secrétariat CITES fournit une copie de ce manuel à toutes les autorités CITES. Si votre institution ne dispose pas d'une copie à jour de ce manuel, veuillez vous adresser au Secrétariat CITES.]

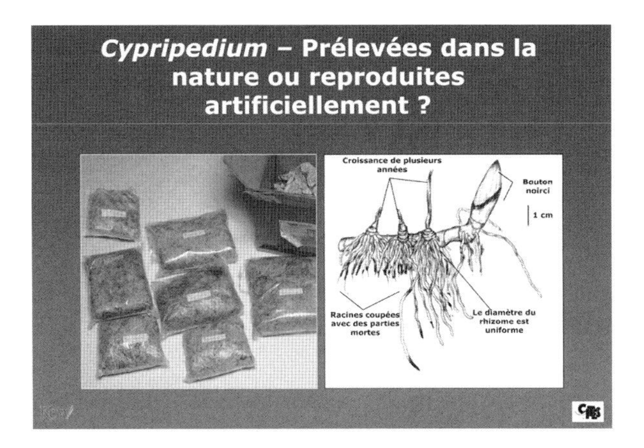

Dia 36 : *Cypripedium* – Prélevées dans la nature ou reproduites artificiellement ?

Il n'est pas facile de distinguer les plantes prélevées dans la nature de celles reproduites artificiellement. Cependant, certaines caractéristiques sont utiles pour faire cette distinction.

Toutes les plantes du genre *Cypripedium* ont un rhizome – une tige souterraine modifiée qui ressemble à une racine.

Le rhizome forme des marques de croissance annuelles qui ressemblent aux perles d'un chapelet. La plupart des espèces ont un rhizome rampant court et robuste qui est rarement ramifié. Chez certaines espèces, telles que *Cypripedium guttatum* et *C. margaritaceum*, le rhizome est allongé et les marques de croissance annuelles sont séparées de quelques centimètres entre elles. Le rhizome survit pendant la période de repos, et le nouveau bourgeon occupe une position terminale. Les véritables racines sont fibreuses et émergent de la partie postérieure de la pousse.

Les *Cypripedium* sont généralement commercialisés au stade dormant. Au printemps ou en automne, ils sont commercialisés sous forme de rhizomes avec des bourgeons et des racines fibreuses. Ce qui semble n'être qu'un sac de racines peut en fait être un grand envoi, même si cela ne ressemble aucunement à ce que le public considère être l'image typique d'une orchidée. Les envois commerciaux de haute qualité sont souvent emballés dans de la tourbe de sphaigne, comme l'illustre l'image de gauche sur la diapositive.

Le rhizome est extrêmement utile, car il peut fournir une indication de l'age de la plante. Il contient les marques de croissance des années précédentes, ce qui permet d'établir un age minimal pour la plante. Si le diamètre du rhizome est constant, cela indique que la plante est mature – par conséquent, elle devrait avoir normalement au moins cinq ans. Chez les plantes immatures, le diamètre du rhizome augmente progressivement jusqu'à atteindre sa taille optimale.

Si vous inspectez un envoi et considérez qu'il a été déclaré comme étant reproduit artificiellement alors qu'il ne l'est pas, vous devriez consulter un expert pour confirmer votre avis.

Dia 37

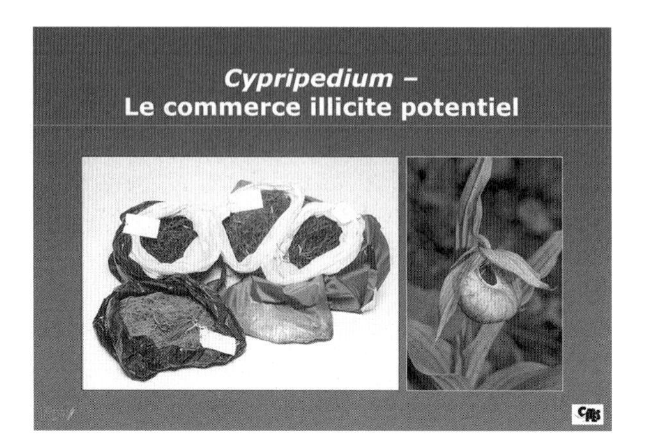

Dia 37 : *Cypripedium* – Le commerce illicite potentiel

Le commerce illicite d'orchidées du genre *Cypripedium* a de fortes chances de concerner les espèces les plus récemment décrites ainsi que celles qui sont difficiles à cultiver. Dans le commerce, ces plantes peuvent être déclarées comme étant reproduites artificiellement. Etant donné que celles-ci sont généralement expédiées sous forme de rhizomes dans les transactions internationales, l'avis d'un expert sera nécessaire pour déterminer si elles ont été prélevées dans la nature ou si elles remplissent la définition de reproduction artificielle de la CITES. Pour vous tenir au courant des espèces récemment décrites, nous vous suggérons de consulter les bases de données mentionnées dans la section de références et ressources. Vérifiez la date où la plante a été scientifiquement décrite pour la première fois. Si elle est récente, il est plus probable que la plante ait été prélevée dans la nature.

[Note pour l'orateur : la diapositive montre des sacs de rhizomes de Cypripedium rhizomes prélevés dans la nature (à gauche) et Cypripedium x froschii (à droite).]

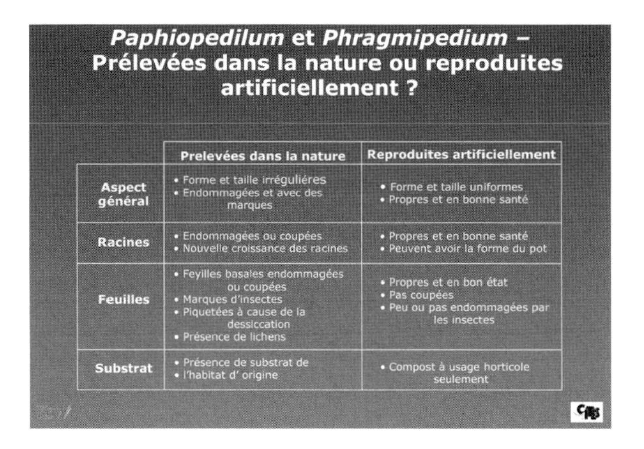

Dia 38 : *Paphiopedilum* et *Phragmipedium* – Prélevées dans la nature ou reproduites artificiellement ?

Cette diapositive passe en revue les caractéristiques clés des plantes des genres *Paphiopedilum* et *Phragmipedium* prélevées dans la nature et de celles reproduites artificiellement. Etant donné que dans le commerce international les orchidées sont généralement commercialisées sans fleurs, ce sera au moyen de l'examen du matériel végétatif qu'il faudra identifier la plante et établir si elle est d'origine sauvage ou non. Tout d'abord, plutôt que de vous concentrer sur l'espèce de laquelle il s'agit, le but est de déterminer si la plante a été prélevée dans la nature. Si la plante remplit un nombre considérable des caractéristiques contenues dans la diapositive et vous pensez qu'elle pourrait avoir été prélevée dans la nature, vous devriez contacter un expert afin qu'il puisse confirmer votre avis.

[Note pour l'orateur : le Comité pour les plantes a élaboré une série de directoires régionaux qui contiennent les coordonnées d'experts en matière de CITES dans les différents pays (visitez le site Web de la CITES pour plus de détails). Vous pourrez utiliser ces informations pour contacter un expert en la matière. Le Volume 1 du Manuel d'identification CITES sur la flore contient des informations sur les caractéristiques des orchidées prélevées dans la nature et de celles reproduites artificiellement. Le Secrétariat CITES fournit une copie de ce manuel à toutes les autorités CITES. Si votre institution ne dispose pas d'une copie à jour de ce manuel, veuillez vous adresser au Secrétariat CITES.]

Dia 39

Dia 39: *Paphiopedilum* et *Phragmipedium* – Prélevées dans la nature ou reproduites artificiellement ?

La diapositive illustre certaines des caractéristiques qui peuvent être présentes chez des plantes prélevées dans la nature ou reproduites artificiellement.

[Note pour l'orateur : la diapositive montre Paphiopedilum spp.]

Dia 40 : *Phragmipedium* – Le commerce illicite potentiel

Dans le cas de ce groupe de plantes, le commerce illicite potentiel se concentrera également sur les espèces récemment décrites. Nous avons souligné le cas de l'espèce *Phragmipedium kovachii*, dont la découverte a éveillé un grand intérêt dans le monde des orchidées et a impulsé le commerce illicite. Cette espèce est toujours très recherchée, et il est probable que le commerce illicite de plantes prélevées dans la nature se poursuivra pendant un certain temps. Il y a de fortes chances que l'espèce fasse l'objet de contrebande sans permis ou qu'elle soit déclarée comme étant reproduite artificiellement sur les permis alors qu'elle ne l'est pas.

Afin de vérifier les noms des espèces de *Phragmipedium* récemment décrites, nous vous proposons de consulter les bases de données mentionnées dans la section de références et ressources.

[Note pour l'orateur : la diapositive montre trois images de Phragmipedium kovachii.]

Dia 41

Dia 41 : *Paphiopedilum* – Le commerce illicite potentiel

Là aussi, le commerce illicite potentiel se concentrera sur les espèces récemment décrites. Les espèces nouvelles les plus intéressantes ont été découvertes en Chine et au Viêt Nam. De nouvelles espèces sont également apparues aux Philippines, en Indonésie et en Malaisie.

Les importations en provenance d'Asie devraient être vérifiées. Outre les envois commerciaux, le plantes peuvent être passées en contrebande dans les bagages enregistrés ou les bagages à main des passagers ou à travers la poste et les services de messagerie.

Afin de vérifier les noms des espèces de *Paphiopedilum* récemment décrites, nous vous proposons de consulter les bases de données mentionnées dans la section de références et ressources. Vous pourrez vérifier la date de publication des noms d'espèces nouvelles – plus cette date sera récente, plus il y aura de chances que la plante en question ait été prélevée dans la nature.

[Note pour l'orateur : la diapositive montre Paphiopedilum micranthum (à gauche) et Paphiopedilum armeniacum (à droite) dans leur habitat d'origine en Chine – espèces étant l'objet de demande de plantes prélevées dans la nature lorsqu'elles sont rentrées dans le commerce dans les années 1980.]

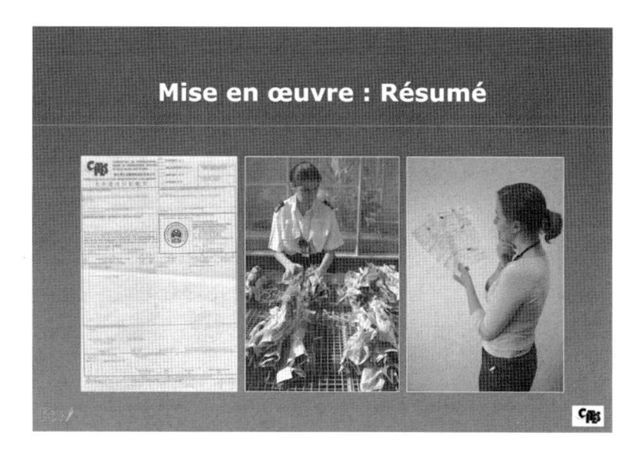

Dia 42 : Mise en œuvre et lutte contre la fraude : Résumé

Nous avons passé en revue les questions clés suivantes sur la mise en œuvre de la CITES pour les sabots de Vénus:

- procédures de mise en œuvre de la Convention et lutte contre la fraude dans différents pays ;
- liste de référence pour les inspections ;
- caractéristiques générales des plantes prélevées dans la nature et de celles reproduites artificiellement ;
- commerce illicite potentiel.

Pour de plus amples informations sur les questions relatives à la lutte contre la fraude et la formation, consultez le site Web de la CITES à l'adresse www.cites.org.

Diapositives supplémentaires

Dia 44

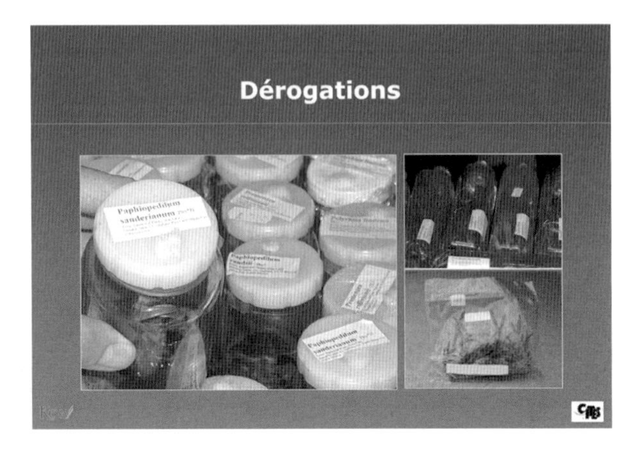

Dia 44 : Dérogations

Il existe bon nombre de plantes dont on ne commercialise que du matériel reproduit artificiellement. En reconnaissant ce fait, les Parties ont décidé d'exempter certains matériels d'orchidées des contrôles de la CITES.

Lorsqu'une espèce est inscrite à l'Annexe I, toute la plante ainsi que ses parties et produits – qu'ils soient vivants ou morts – sont contrôlés. Il n'y a qu'une exception : les cultures de plantules ou de tissus « obtenus *in vitro* en milieu solide ou liquide et transportées en conteneurs stériles » sont exclues des contrôles. Naturellement, le matériel en question doit être d'origine licite pour faire l'objet de cette dérogation.

Lorsqu'une espèce est inscrite à l'Annexe II de la CITES, la plante « vivante ou morte » est contrôlée ainsi que toute partie ou tout produit facilement identifiable mentionné dans les annexes. Dans le cas de *Cypripedium*, les seuls parties et produits exclus des contrôles de la CITES sont : a) les graines et le pollen (y compris les pollinies) ; b) les cultures de plantules ou de tissus obtenues *in vitro* en milieu solide ou liquide et transportées en conteneurs stériles ; ainsi que c) les fleurs coupées de plantes reproduites artificiellement.

[Note pour l'orateur : quoique bon nombre de taxons hybrides d'orchidées aient été exclus des contrôles de la CITES, ces dérogations ne touchent pas les sabots de Vénus. Pour obtenir plus de détails sur les dérogations qui s'appliquent à d'autres hybrides d'orchidées, consultez le site Web de la CITES à l'adresse www.cites.org.]

Dia 45 : Le système d'enregistrement des pépinières de la CITES

Les procédures établies pour l'enregistrement des pépinières sont stipulées dans la Résolution Conf 9.19 (Rev. CoP13), *Lignes directrices pour l'enregistrement des pépinières exportant des spécimens reproduits artificiellement d'espèces inscrites à l'Annexe I.* Cette résolution a été adoptée à la 9e réunion de la Conférence des Parties à Fort Lauderdale, Etats-Unis en novembre 1994, et révisée à la CdP13 à Bangkok en 2004. La Convention n'a pas établi de critères pour l'enregistrement de pépinières qui cultivent des plantes de l'Annexe II. Cependant, tous les organes nationaux CITES sont libres d'établir un système d'enregistrement pour les plantes de l'Annexe II avec une procédure simple de délivrance des permis. Cela bénéficierait les organes et les commerçants locaux. Toutefois, l'enregistrement ne serait pas reconnu en dehors du pays en question.

L'organe de gestion de toute Partie peut, en consultation avec son autorité scientifique, présenter une demande pour qu'une pépinière reproduisant des espèces inscrites à l'Annexe I soit inscrite au registre du Secrétariat CITES. Le propriétaire de la pépinière doit d'abord fournir des informations sur son établissement à l'organe de gestion national. Ces informations devraient comprendre, entre autres, une description des installations, une description des antécédents de la pépinière et de ses projets en ce qui concerne la reproduction artificielle de plantes, et la description de la population parentale – quantité et type de plantes de l'Annexe I en existence en apportant la preuve de son acquisition licite. L'organe de gestion et l'autorité scientifique doivent étudier ces informations et juger si l'établissement remplit les conditions pour être inscrit au registre. Dans cette démarche, ce serait normal que les autorités nationales réalisent une inspection assez minutieuse de la pépinière.

Lorsque les organes CITES nationaux jugent que la pépinière remplit toutes les conditions d'enregistrement, elles transmettent leur avis et les informations sur la pépinière au Secrétariat CITES. L'organe de gestion doit également décrire les procédures d'inspection suivies pour confirmer l'identité et l'origine licite de la population parentale des plantes qui vont faire partie du système de registre et de tout autre matériel de l'Annexe I. Les organes CITES nationaux doivent aussi garantir que toute population parentale d'origine sauvage n'est pas appauvrie et que l'établissement fait l'objet d'un suivi détaillé. Si l'établissement utilise des graines prélevées dans la nature, l'organe de gestion devrait certifier que les conditions mentionnées dans la définition de reproduction artificielle de la CITES sont respectées (voir Dia 46). L'organe de gestion CITES devrait également mettre en place une procédure simple de délivrance des permis et en informer le Secrétariat dans le détail.

Si le Secrétariat CITES est satisfait avec les informations fournies, il doit inscrire la pépinière dans son registre. Lorsqu'une pépinière ne remplit pas toutes les conditions d'enregistrement, le Secrétariat doit fournir à l'organe de gestion une explication complète et indiquer les conditions spécifiques à remplir. Tout organe de gestion ou toute autre source peut informer le Secrétariat de toute irrégularité en ce qui concerne les conditions d'enregistrement. Si ces problèmes ne sont pas résolus, la pépinière peut être supprimée du registre après consultation de l'organe de gestion national.

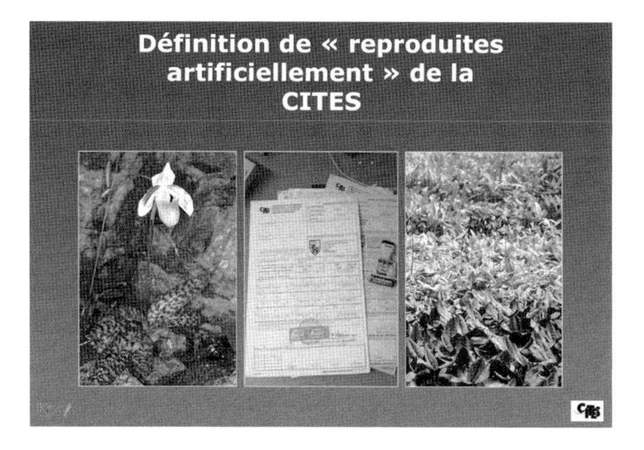

Dia 46 : Définition de « reproduites artificiellement » de la CITES

La définition de « reproduites artificiellement » de la CITES figure dans la Résolution Conf. 11.11 (Rev. CoP13) – *Réglementation du commerce des plantes*. La définition de la CITES comprend plusieurs critères uniques. L'application de ces critères peut donner lieu à ce qu'une plante qui présente toutes les caractéristiques physiques relevant de la reproduction artificielle soit considérée comme prélevée dans la nature pour la Convention. Voici les points clés :

- *Les plantes doivent pousser dans des conditions contrôlées*. Cela veut dire que les plantes sont manipulées dans un *milieu non naturel* pour encourager les meilleures conditions de croissance et exclure les prédateurs. Une pépinière traditionnelle ou une simple serre sont des « conditions contrôlées ». Une installation visant à protéger les plantes de l'excès de soleil serait également un exemple de « conditions contrôlées ». Par contre, l'ajout provisoire d'un morceau de végétation naturelle où l'on trouve déjà des spécimens sauvages de plantes ne serait pas considéré comme des « conditions contrôlées ». En outre, les plantes prélevées dans la nature sont considérées sauvages même si elles ont été cultivées dans des conditions contrôlées pendant un certain temps.

- La population parentale cultivée doit être *établie de manière non préjudiciable à la survie de l'espèce dans la nature*, et gérée de manière à *garantir le maintien à long terme de cette population parentale*.

- La population parentale cultivée doit être *établie conformément aux dispositions de la CITES et aux lois nationales pertinentes*. Cela signifie que la population parentale doit être obtenue légalement du point de vue de la CITES mais également de toute la législation nationale du pays d'origine. Par exemple, une plante peut avoir été prélevée illégalement dans son pays d'origine, cultivée ensuite dans une pépinière locale, et sa progéniture peut être déclarée comme reproduite artificiellement pour être exportée. Pour la CITES, cette progéniture ne peut pas être considérée reproduite artificiellement en raison du prélèvement illégal des plantes parentales.

Dia 46

- Les graines ne sont considérées comme reproduites artificiellement que si elles sont issues de spécimens qui correspondent à la définition de « reproduites artificiellement » de la CITES. Le terme *population parentale cultivée* est utilisé pour permettre l'ajout de nouvelles plantes prélevées dans la nature à la population parentale. On reconnaît que de temps en temps il peut être nécessaire d'ajouter des plantes sauvages à la population parentale, ce qui est permis à condition que cela soit fait de manière légale et durable.

- Lorsque les plantes et les graines sont issues de graines prélevées dans la nature et cultivées dans un Etat de l'aire de répartition, elles peuvent être considérées reproduites artificiellement si cela est autorisé par l'organe de gestion et l'autorité scientifique du pays en question.

Ce n'est pas facile d'appliquer la définition de la CITES. Il faut vérifier l'acquisition licite des plantes, les conditions de reproduction et le prélèvement non préjudiciable des spécimens. Afin de pouvoir y parvenir, l'organe de gestion et l'autorité scientifique CITES doivent travailler en collaboration étroite. L'application quotidienne des critères doit être adaptée à la situation de chaque Partie. Les autorités CITES nationales devraient envisager d'élaborer une liste de référence afin de normaliser la démarche et d'informer les commerçants de plantes locaux.

Dia 47 : Promotion du commerce durable et accès au matériel de reproduction

L'inscription d'espèces végétales et animales à l'Annexe I de la CITES représente l'interdiction des transactions commerciales pour ces espèces. Le but de cette mesure est de protéger les espèces en question d'un commerce préjudiciable qui pourrait les mener à l'extinction. En soi, l'inscription d'une espèce à l'Annexe I ne devrait pas être considérée comme un succès en matière de conservation. Au contraire, c'est plutôt lorsque l'espèce en question peut être transférée à l'Annexe II qu'il faut s'en réjouir du point de vue de la conservation. Par conséquent, après l'inscription d'une espèce à l'Annexe I de la CITES, il est important de mettre en œuvre des mesures pour la conservation de l'espèce. La demande d'espèces ne disparaît pas lorsque celles-ci sont inscrites à l'Annexe I, mais en principe il devrait y avoir du matériel cultivé disponible pour satisfaire cette demande. Ceci est possible dans les cas où l'on a développé des techniques adéquates pour la reproduction artificielle et que des plantes parentales d'origine licite sont disponibles.

Cependant, dans le cas des sabots de Vénus, de nouvelles espèces sont continuellement recherchées, trouvées et décrites. Il est souvent difficile d'obtenir des plantes parentales licites pour la reproduction de ces espèces. Dans ces cas, le matériel illégal se filtre dans le marché international et s'incorpore progressivement au matériel de reproduction. Cette démarche encourage le prélèvement non durable des espèces les plus rares et prive les pays d'origine des revenus considérables qui pourraient découler de l'introduction de tels matériels reproducteurs dans le marché international. Les pays d'origine, le commerce international d'orchidées et la communauté CITES ont fait preuve d'une certaine inertie. Nous attendons une coopération réussie pour établir les mécanismes qui permettront l'accès au matériel de reproduction tout en aidant à vaincre le commerce illicite. Pour mettre en place de tels programmes, les pays d'origine ont besoin d'assistance.

Au sein de la CITES, il est possible de mettre en place des initiatives pour que cela puisse se produire. Seuls sont nécessaires de l'enthousiasme, de l'initiative, de la confiance et du financement. La confiance est probablement l'ingrédient le plus difficile à obtenir. De telles initiatives visant à la reproduction seront toujours vulnérables au commerce illicite ou aux accusations de « biopiraterie ». Si vous travaillez dans le contexte de la CITES ou dans le commerce des orchidées, vous devriez essayer de promouvoir de telles initiatives. Ce type de projets encourage le commerce durable et permet aux pays d'origine d'avoir accès aux revenus générés par leurs propres ressources. Il se pourrait qu'ils représentent également la seule manière de consolider les partenariats nécessaires pour la conservation à long terme des espèces et de leurs habitats.

INDEX

CITES y las zapatillas de Venus

Una introducción a las zapatillas de Venus amparadas por la Convención sobre el Comercio Internacional de Especies Amenazadas

Autores

H. Noel McGough,

David L. Roberts, Chris Brodie y Jenny Kowalczyk

Royal Botanic Gardens, Kew

Reino Unido

Consejo de Administración, Royal Botanic Gardens, Kew

2006

ÍNDICE

INTRODUCCIÓN

"*CITES y las zapatillas de Venus*" se ha producido con el objetivo de ofrecer una introducción a las zapatillas de Venus en CITES. Incluye la identificación, el comercio, y la aplicación de la Convención para estas orquídeas.

La principal intención de la guía es que sirva como un instrumento de formación para los que trabajan con la Convención, a saber, Autoridades Administrativas, Autoridades Científicas y organismos encargados del cumplimiento de CITES. Sin embargo, probablemente sea útil también para un público mucho más amplio, especialmente aquellas personas que deseen aprender sobre el funcionamiento de CITES con respecto a este grupo vegetal de importancia comercial.

"*CITES y las zapatillas de Venus*" se ha diseñado para poder adaptarse fácilmente a las necesidades del ponente. Alentamos al usuario a que realice cuantos "ajustes" estime convenientes para aproximarse a su público. Además de los apuntes para ponentes, proporcionamos una bibliografía y una lista de recursos. Esperamos que Ud. encuentre el paquete útil, no solo para desarrollar sus presentaciones, sino también como una obra de referencia práctica. Por favor, utilice este instrumento didáctico, y envíenos sus comentarios para que podamos revisar ediciones posteriores y adecuarlas a sus necesidades.

Noel McGough,

Director, Sección de Convenciones y Política,

Autoridad Científica CITES del Reino Unido para Plantas,

Royal Botanic Gardens, Kew

AGRADECIMIENTOS

Los autores quisieran agradecer a las siguientes personas el asesoramiento técnico prestado durante la preparación de este paquete: Wendy Byrnes, Phillip Cribb, Margarita Clemente Muñoz, Deborah Rhoads Lyon, Matthew Smith, Sabina Michnowicz, y Ger van Vliet.

El paquete ha sido financiado por la Autoridad Administrativa CITES del Reino Unido, el Departamento de Medioambiente, Alimentación y Asuntos Rurales (Defra).

CÓMO USAR ESTE PAQUETE-PRESENTACIÓN

Este paquete consta de diapositivas y apuntes para los ponentes de una presentación sobre las zapatillas de Venus incluidas en los Apéndices CITES. La presentación se compone de tres áreas temáticas independientes que pueden utilizarse y adaptarse según los intereses, las características y necesidades particulares de su público (Introducción a las zapatillas de Venus, Zapatillas de Venus en CITES, Aplicación de CITES para zapatillas de Venus).

Un cuarto apartado de diapositivas y apuntes amplía detalles sobre algunos temas más, que pueden añadirse a la presentación si se estima conveniente. Las diapositivas se han redactado en términos generales, con la esperanza de que sigan siendo de actualidad, y por tanto útiles, durante un futuro previsible.

Cada diapositiva viene acompañada de unos apuntes, a modo de sugerencias para ponentes. Dichos apuntes son más específicos que las diapositivas y reflejan datos actualizados hasta mayo de 2005. Por supuesto se alienta a los ponentes que expresen su estilo personal; no es necesario seguir los apuntes al pie de la letra. ¡Hagan el uso de ellos que crean más cómodo!

Esperamos que este paquete sirva de punto de partida útil, desde donde adaptar las diapositivas y apuntes correspondientes para reflejar las necesidades concretas de su público, la duración de la presentación, y su propio estilo personal. Por ejemplo, UD. podría ilustrar algunas diapositivas con ejemplos de su propia región o institución, o complementarlas con imágenes adicionales como cómics, fotos, o recortes de periódicos. Así, indudablemente aumentará el impacto de una presentación individual. Además, las diapositivas pueden imprimirse, a partir del archivo de Microsoft® PowerPoint® del CD-ROM, en hojas de transparencias, para presentarlas con un retro-proyector o repartirlas como información al público.

CD-ROM

El CD-ROM contiene los siguientes archivos:

- "CITESSlipperOrchids.ppt", una presentación Microsoft PowerPoint® con las diapositivas y los apuntes para ponentes. Ud. deberá tener Microsoft PowerPoint 97® (o una versión más reciente) instalado en su ordenador para ver y adaptar este archivo.

- "CITESSlipperOrchids.pdf", una presentación Adobe Acrobat®. No podrá realizar modificaciones sobre este formato, pero se puede ver "a pantalla completa" con Adobe Reader®. Deberá tener Adobe Acrobat Reader® instalado en su ordenador para ver este archivo (puede descargarse de www.adobe.com).

- "CITESSlipperOrchidsBW.pdf", una presentación Adobe Acrobat® en blanco y negro (puede descargarse de www.adobe.com).

- "CITESSlipperOrchidsPack.pdf", copia completa del texto del paquete, con su introducción, referencias y apuntes para ponentes. Permite ver el documento electrónico entero, además de imprimirlo en parte o en su totalidad. Deberá tener Adobe Acrobat Reader® instalado en su ordenador para ver este archivo (puede descargarse de www.adobe.com).

REFERENCIAS Y RECURSOS

Referencias a la Convención

CITES (2003 y actualizaciones). *CITES Handbook.* Secretaría de la Convención sobre el Comercio Internacional de Especies de Fauna y Flora Silvestres. Ginebra, Suiza. Este manual incluye el texto de la Convención y sus Apéndices, una copia de un permiso normalizado, y el texto de las Resoluciones y Decisiones de la Conferencia de las Partes.

Wijnstekers, W. (2003 y actualizaciones). *The Evolution of CITES, 6th edition.* Secretaría de la Convención sobre el Comercio Internacional de Especies de fauna y flora Silvestres. Ginebra, Suiza. La referencia más exhaustiva disponible, y la primera autoridad sobre el Convenio. Escrito por el Secretario General de CITES. Se actualiza con regularidad.

Rosser, A. y Haywood, M. (Recopiladores), (2002). *Guidance for CITES Scientific Authorities. Checklist to assist in making non-detriment findings of Appendix II exports.* Occasional Paper of the IUCN Species Survival Commission No. 27. UICN-Unión Mundial para la Naturaleza, Gland, Suiza y Cambridge, Reino Unido. El primer intento de definir directrices para el uso de las Autoridades Científicas al hacer dictámenes de extracciones no perjudiciales del medio silvestre; dictámenes que son el requisito previo a la concesión de permisos de exportación CITES.

La página web de CITES (www.cites.org) contiene una amplia gama de información sobre el Convenio, las especies incluidas en los Apéndices, direcciones y contactos de interés, informes de reuniones y grupos de trabajo, publicaciones y páginas web nuevas, y un calendario de acontecimientos.

Críticas de la Convención

Hutton, J. y Dickson, B. (2000). *Endangered Species, Threatened Convention. The Past, Present and Future of CITES.* Earthscan, Londres, Reino Unido. Una evaluación crítica de CITES desde la perspectiva del uso sostenible.

Oldfield, S. (Editor), (2003). *The Trade in Wildlife: Regulation for Conservation.* Earthscan, Londres, Reino Unido. Una mirada crítica al comercio internacional de especies silvestres.

Reeve, R. (2002). *Policing International Trade in Endangered Species. The CITES Treaty and Compliance.* Royal Institute of International Affairs. Earthscan, Londres, Reino Unido. Un estudio detallado del sistema de cumplimiento de CITES.

Referencias normalizadas de CITES para plantas – Listas de referencia

Carter, S. y Eggli, U. (2003). *The CITES Checklist of Succulent Euphorbia Taxa (Euphorbiaceae).* Segunda edición. German Federal Agency for Nature Conservation, Bonn, Alemania. Referencias a los nombres de especies suculentas de *Euphorbia.*

Hunt, D. (1999). *CITES Cactaceae Checklist.* Segunda edición. Royal Botanic Gardens, Kew, Reino Unido. Referencias a los nombres de Cactaceae, la familia de los cactus.

Mabberley, D.J. (1997). *The Plant-Book.* Segunda edición. Cambridge University Press, Cambridge, Reino Unido. La referencia para la denominación genérica de todas las especies vegetales CITES, hasta que se sustituya por las listas normalizadas adoptadas por la Conferencia de las Partes como se indica en esta relación.

Newton, L.E. y Rowley, G.D. (Eggli, U. Editor), (2001). *CITES Aloe and Pachypodium Checklist.* Royal Botanic Gardens, Kew, Reino Unido. Referencia para nombres de *Aloe* y *Pachypodium.*

Roberts, J.A., Beale, C.R., Benseler, J.C., McGough, H.N. y Zappi, D.C. (1995). *CITES Orchid Checklist. Volume 1.* Royal Botanic Gardens, Kew, Reino Unido. Referencias a los nombres de *Cattleya, Cypripedium, Laelia, Paphiopedilum, Phalaenopsis, Phragmipedium, Pleione* y *Sophronitis* con reseñas sobre *Constantia, Paraphalaenopsis* y *Sophronitella.*

Roberts, J.A., Allman, L.R., Beale, C.R., Butter, R.W., Crook, K.B. y McGough, H.N. (1997). *CITES Orchid Checklist. Volume 2.* Royal Botanic Gardens, Kew, Reino Unido. Referencias a los nombres de *Cymbidium, Dendrobium, Disa, Dracula* y *Encyclia.*

Roberts, J.A., Anuku, A., Burdon, J. , Mathew, P., McGough, H.N. y Newman, A.D. (2001). *CITES Orchid Checklist. Volume 3.* Royal Botanic Gardens, Kew, Reino Unido. Referencias a los nombres de *Aerangis, Angraecum, Ascocentrum, Bletilla, Brassavola, Calanthe, Catasetum, Miltonia, Miltonioides, Miltoniopsis, Renanthera, Renantherella, Rhynchostylis, Rossioglossum, Vanda* y *Vandopsis.*

Willis, J.C., revisado por Airy Shaw, H.K. (1973). *A Dictionary of Flowering Plants and Ferns.* 8ª edición. Cambridge University Press. Cambridge, Reino Unido. Para los sinónimos genéricos que no figuran en *The Plant-Book*, hasta que sean sustituidos por los de las listas normalizadas adoptadas por la Conferencia de las Partes como se indica en esta relación.

UNEP-WCMC (2005). *Checklist of CITES Species.* PNUMA-Centro Mundial de Monitoreo de la Conservación, Cambridge, Reino Unido. La CdP ha adoptado esta Lista de referencia y sus actualizaciones como compendio oficial de los nombres científicos que se encuentran en las referencias normalizadas.

El Comité de Nomenclatura actualiza las Listas de referencia CITES con regularidad. Véase la página web de CITES para más información: www.cites.org.

Referencias adicionales

Las siguientes son referencias generales que esperamos sean de utilidad. Sepan que la taxonomía en estas obras puede ser distinta a la indicada en las referencias adoptadas por CITES señaladas arriba. Por favor, infórmenos Ud. de obras que encuentre útil y las incluiremos en ediciones futuras de la guía.

Averyanov, L., Cribb, P., Loc P.K. & Hiep. N.H. (2003). *Slipper Orchids of Vietnam.* Royal Botanic Gardens, Kew. Reino Unido. Amplio tratado con descripciones completas de todas las especies de *Paphiopedilum* autóctonas de Vietnam, dibujos minuciosos y uso extensivo de fotos en color, incluidas algunas del hábitat.

Referencias y recursos

Bechtel, H. Cribb, P. y Launert, E. (1992). *The Manual of Cultivated Orchid Species*. Tercera edición. Blanford Press, Londres, Reino Unido. Debería ser re-editada, pero aún constituye una referencia excelente, con un análisis detallado de más de 400 géneros, 1200 especies, más de 860 fotos en color y muchos dibujos muy minuciosos.

Braem, G. J., Baker, C.O y Baker, M.L. (1998). *The Genus Paphiopedilum: Natural History and Cultivation, Volume 1.* Botanical Publishers, Inc., Kissimmee, Florida, EE.UU.

Braem, G. J., Baker, C.O y Baker, M.L. (1999). *The Genus Paphiopedilum: Natural History and Cultivation, Volume 2.* Botanical Publishers, Inc., Kissimmee, Florida, EE.UU.

Braem, G. J. y Chiron, G.R. (2003). *Paphiopedilum.* Tropicalia, Saint-Genis Laval, Francia.

Cash, C. (1991). *The Slipper Orchids.* Christoper Helm, Londres, Reino Unido.

Cavestro, W. (2001). *Le genre Paphiopedilum: taxonomie, répartition, habitat, hybridation et culture.* Rhône-Alpes Orchidées, Lyon, Francia.

Chen, V.Y. y Song, M. (2000). *Guide to CITES Plants in Trade.* (Edición china). TRAFFIC East Asia.

(CITES (1993-). *CITES Identification Manual, Volume 1 Flora.* Secretaría de la Convención sobre el Comercio Internacional de Especies de fauna y flora Silvestres. Ginebra, Suiza. Éste es el manual de identificación oficial de CITES. Las Partes se comprometen a elaborar hojas para el manual si prospera su propuesta de inclusión de una especie en los Apéndices. Se trata de un cuaderno que se puede abrir con anillas, añadiendo nuevas hojas de identificación constantemente. Imprescindible para cualquiera que trabaje con CITES y las plantas.

Comisión Europea (2002). *"La nueva normativa sobre comercio de especies silvestres cumple cinco años".* En el sitio Internet de la Dirección General (DG) de Medio Ambiente (página sólo disponible en inglés): http://europa.eu.int/comm/environment/cites/info_en.htm. Una vez que haya accedido a dicha dirección, debe ir hasta la publicación titulada "1997-2002: Five years of new wildlife trade regulations" y está en inglés "en" y en español "es".

Cribb, P. (1997). *Slipper Orchids of Borneo.* Natural History Publications (Borneo), Kota Kinabalu.

Cribb, P. (1997). *The Genus Cypripedium – A Botanical Magazine Monograph.* Publicado en asociación con Royal Botanic Gardens, Kew. Timber Press, Portland, EE.UU. Estudio monográfico exhaustivo con una completa descripción de todas las especies, fotos en color y minuciosos dibujos en negro y en color.

Cribb, P. (1998). *The Genus Paphiopedilum (Segunda edición) – A Botanical Magazine Monograph.* Publicado en asociación con Royal Botanic Gardens, Kew. Natural History Publications (Borneo), Kota Kinabalu. Estudio monográfico exhaustivo con una completa descripción de todas las especies, fotos en color y minuciosos dibujos en negro y en color.

Grupo de Especialistas de Orquídeas del UICN/CSE. (1996). *Orchids – Status Survey and Conservation Action Plan.* UICN, Gland Suiza y Cambridge, Reino Unido.

Gruss, O. (2003). *A Checklist of the Genus Phragmipedium.* Orchid Digest. 67[4]: 213-255.

Hennessy, E. F. y Hedge, T.A. (1989). *The Slipper Orchids.* Acorn Books, Randburg, República de Sudáfrica.

Hilton-Taylor, C. (Recopilador), (2000-). *IUCN Red List of Threatened Species.* UICN-Unión Mundial para la Naturaleza. Gland, Suiza y Cambridge, Reino Unido. La lista oficial de la UICN de plantas y animales amenazadas, publicada como folleto con CD-ROM. Se actualiza y mejora constantemente; para la última versión, visite la página web de la Lista Roja en www.redlist.org.

Jenkins, M. and Oldfield, S. (1992). *Wild Plants in Trade.* TRAFFIC International, Cambridge, Reino Unido. Resumen del último estudio completo del comercio europeo de especies vegetales CITES.

Koopowitz, H. (2000). *A revised checklist of the Genus Paphiopedilum.* Orchid Digest. 64[4]: 155-179.

Lange, D. y Schippmann, U. (1999). *Checklist of medicinal and aromatic plants and their trade names covered by CITES and EU Regulation 2307/98 Version 3.0.* German Federal Agency for Nature Conservation. Bonn, Alemania.

Marshall, N.T. (1993). *The Gardener's Guide to Plant Conservation.* TRAFFIC North America. Desgraciadamente ya se ha quedado anticuada, aunque hay rumores de una próxima reedición. Era una guía muy útil sobre el comercio con plantas para la horticultura y los orígenes de las mismas.

Mathew, B. (1994). *CITES Guide to Plants in Trade.* UK Department of the Environment, Londres, Reino Unido. Ya anticuada, pero tiene fotos en color y descripciones de los grupos de plantas más importantes de CITES controlados y comercializados a principios de los 1990.

McCook, L. (1998). *An annotated checklist of the genus Phragmipedium.* Orchid Digest Corp. California, EE.UU. Publicación especial del *Orchid Digest.*

Pridgeon, A. (2003). *The Illustrated Encyclopedia of Orchids.* David y Charles, Devon, Reino Unido. Más de 1100 especies ilustradas. Repasa los principales taxa objeto de comercio y de interés para coleccionistas. Fotos a todo color. La mejor guía general de orquídeas impresa.

Rittershausen, W. & B. (1999). *Orchids – a practical guide to the world's most fascinating plants.* The Royal Horticultural Society, reimpreso 2004. Quadrille Publishing Ltd, Londres, Reino Unido.

Sandison, M. S., Clemente Muñoz, M., de Koning J. y Sajeva, M. (1999). *CITES and Plants – A User's Guide.* Royal Botanic Gardens, Kew, Reino Unido. Primer "paquete de diapositivas" de 40 diapositivas y texto producido en inglés, francés y castellano.

Sandison, M. S., Clemente Muñoz, M., de Koning J. y Sajeva, M. (2000). *CITES and Plants – A User's Guide.* (Chinese Edition). Royal Botanic Gardens, Kew,

Reino Unido. Editado por Vincent Y. Chen y Michael Song; producido por TRAFFIC East Asia. La versión original de la Guía del usuario en lengua china.

Schippmann, U. (2001). *Medicinal Plants Significant Trade Study CITES Project S-109. Plants Committee Document PC9 9.1.3.(rev.). BfN - Skripten 39.* German Federal Agency for Nature Conservation, Bonn, Alemania. Una excelente visión global de la conservación y del comercio con plantas medicinales incluidas en CITES.

CD-ROM

CITES (2002-). *CITES training presentations.* Secretaría CITES, Ginebra, Suiza. Una serie de presentaciones formativas producidas por la Dependencia de Creación de Capacidades de la Secretaría CITES. En formato CD-ROM de tamaño "tarjeta de visita", son instrumentos imprescindibles para las personas que trabajan en la formación sobre CITES.

CITES (2003-). *CD-ROM version of the CITES website* (*www.cites.org*). Versión completa de la página web de la CITES en CD-ROM. Se puede pedir de la Secretaría CITES.

Páginas Web

Existen numerosas páginas web de cierto interés para las personas que trabajan con CITES. Muchas Autoridades nacionales le han dedicado un sitio web. A continuación sugerimos unos sitios clave que pueden llevarle a Ud. a cuantas otras páginas como tenga Ud. tiempo para navegar en la Red.

CITES, Página de Inicio: Sitio oficial de la Secretaría CITES. Incluye listas de las Partes, las Resoluciones y otros documentos. www.cites.org.

Comisión Europea: Información respecto de la normativa de aplicación sobre comercio con especies silvestres amparadas por CITES dentro de la Unión Europea. www.eu-wildlifetrade.org.

Sitio web CITES del Reino Unido: Una página web mantenida por las Autoridades CITES del RU, con el objetivo de proporcionar información actualizada sobre temas relacionados con CITES en lo que afecten al Reino Unido y sus dependencias en ultramar. www.ukcites.gov.uk.

UICN-Unión Mundial para la Naturaleza: La mayor organización profesional para la conservación. La UICN reúne a administraciones, organizaciones no-gubernamentales, instituciones e individuos para ayudar a las naciones a sacar el mejor provecho de sus recursos naturales de manera sostenible. www.iucn.org.

Comisión de Supervivencia de Especies de la UICN: La CSE es la principal fuente de información científica y técnica que tiene la UICN para la conservación de especies amenazadas y vulnerables de flora y fauna. Realiza tareas concretas en nombre de la UICN, como el control de las especies vulnerables y sus poblaciones, la aplicación y el examen de los planes de acción para la conservación. Da directrices, consejos y recomendaciones políticas a los gobiernos, las agencias y organizaciones sobre la conservación y gestión de las especies y de sus poblaciones. www.iucn.org/themes/ssc/.

PNUMA-Centro Mundial de Monitoreo de la Conservación: (UNEP-WCMC) realiza servicios de información sobre la conservación y uso sostenible de los recursos vivos del planeta, ayudando a otros a confeccionar sistemas de información. Sus actividades incluyen el apoyo a la Secretaría CITES. Se pueden pedir datos sobre el comercio internacional de especies silvestres y estadísticas comerciales del Programa de Especies del UNEP - WCMC. Ahora es una oficina de la ONU basada en Cambridge, Reino Unido, pero el trabajo del Centro forma parte íntegra del Programa de las Naciones Unidas para el Medio Ambiente (PNUMA), con sede en Nairobi, Kenia. www.unep-wcmc.org/index.html.

TRAFFIC International: TRAFFIC es un programa del WWF (Fondo Mundial para la Naturaleza) y de la UICN establecido para controlar el comercio con plantas y animales silvestres. La Red de TRAFFIC es el mayor programa de vigilancia mundial, con oficinas que abarcan la mayor parte del mundo. Trabaja estrechamente con la Secretaría CITES. www.traffic.org.

Earth Negotiations Bulletin: Sigue las negociaciones medioambientales más importantes a medida que ocurran. También dispone de material de archivo extenso y muchas fotos de las reuniones. www.iisd.ca.

Verificación de nombres de plantas

Los siguientes sitios web son útiles para verificar nombres de plantas que no se encuentran en las listas de referencia normalizadas de la CITES. A veces estos nombres pueden corresponder a especies de reciente descripción. Si este "nombre nuevo" se ha utilizado en una solicitud de un permiso CITES declarando que la planta se ha reproducido artificialmente, hay que comprobar la identidad de la planta y asegurar que cumpla la definición CITES de reproducción artificial.

IPNI – The International Plant Names Index: Una base de datos, con los nombres y detalles bibliográficos correspondientes, de todos los espermatófitos. www.ipni.org/index.html.

TROPICOS: Una base de datos de nomenclatura producida y mantenida por el Missouri Botanical Garden. mobot.mobot.org/W3T/Search/vast.html.

EPIC – Electronic Plant Information Centre: Reúne toda la información digitalizada sobre las plantas en posesión de los Royal Botanic Gardens, Kew. www.rbgkew.org.uk/epic/.

World Checklist of Monocotyledons – Constituye un inventario de los nombres taxonómicamente validados de las plantas monocotiledóneas con los detalles bibliográficos pertinentes, además de su distribución global. Incluye una lista completa de todos los nombres de orquídeas. www.kew.org/monocotChecklist/

Phragweb – Una fuente exhaustiva de información sobre especies de *Phragmipedium*, con descripciones, dibujos minuciosos y fotos a todo color.
www.Phragweb.info.

RHS – La página web de la Royal Horticultural Society, útil para comprobar nuevos nombres de híbridos.
www.rhs.org.uk/publications/pubs_journals_orchid_hybrid.asp.

ÍNDICE DES DIAPOSITIVAS

Diapositivas de introducción

Introducción a las zapatillas de Venus

Zapatillas de Venus en CITES

Aplicación de CITES para zapatillas de Venus

Diapositivas adicionales

Diapositiva 1: La CITES y las zapatillas de Venus

El objetivo de esta presentación es ofrecer al usuario una introducción a los diferentes tipos de zapatillas de Venus amparadas por la Convención sobre el Comercio Internacional de Especies Amenazadas de Fauna y Flora Silvestres – CITES – y comentar algunos de los principales temas relacionados con la aplicación de la Convención para este importante grupo vegetal.

Diapositiva 2

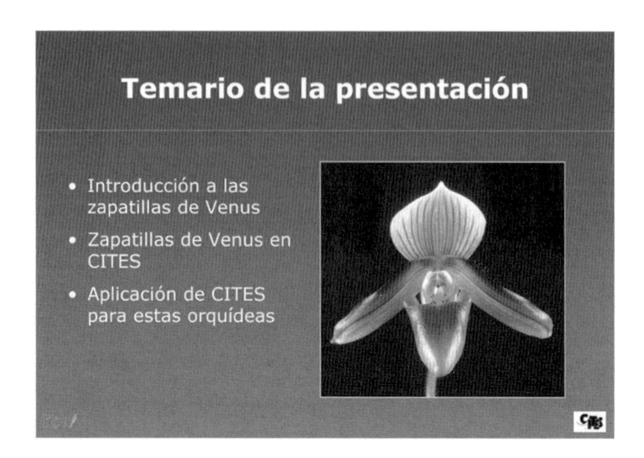

Diapositiva 2: Temario de la presentación

Esta presentación abarcará los siguientes temas:

- Introducción a las zapatillas de Venus;
- Zapatillas de Venus amparadas por CITES;
- Aplicación de la Convención para estas orquídeas.

[Nota al ponente: La diapositiva muestra Paphiopedilum callosum.]

Diapositiva 3: Diversidad de orquídeas

La mayoría de la gente tiene al menos cierta idea de lo que es una orquídea. ¡El nombre "orquídea" a menudo nos encandila con ideas de flores fascinantes, ambientes exóticos y regalos impresionantes! Sin embargo, las plantas que ve la mayoría de la gente representan tan sólo una minúscula porción de lo que es posiblemente el mayor grupo vegetal del mundo, con más de 25.000 especies conocidas, y se calcula que pueden existir otras 5.000 especies aún sin descubrir. Estas especies se encuentran en todo el mundo, con la mayor parte (alrededor del 70%) concentrada en las selvas tropicales, aunque las orquídeas se pueden dar hasta en zonas muy áridas y subárticas, como Alaska. ¡Se han encontrado incluso especies subterráneas en Australia! Su tamaño puede variar desde el de la punta de un lápiz hasta el de un vigoroso arbusto de dos toneladas. Además, al contrario de lo que se suele creer, no todas las orquídeas son raras. En el hábitat adecuado, algunas poblaciones pueden ser muy numerosas. Por su tremenda variedad de tamaños, colores, y su atractivo exótico, las orquídeas intrigan a coleccionistas y viveristas desde mediados del siglo XIX.

[Nota al ponente: Las características que distinguen a las orquídeas de otras plantas se limitan principalmente a las flores, y son las siguientes:

1. *Las partes macho y hembra de la flor se funden, al menos parcialmente, en una estructura llamada la columna,*

2. *Uno de los pétalos de la flor se suele presentar muy modificado, actuando como una pista de aterrizaje o guía para los insectos polinizadores; se llama el labelo o labio,*

3. *El polen normalmente se encuentra en grandes masas llamadas polinias, por pares o en grupos de 4, 6, u 8,*

4. *Las orquídeas producen millones de semillas minúsculas que no tienen ninguna fuente de alimentación para germinar, por lo que tienen que asociarse con un hongo para estimular la germinación.*

La diapositiva muestra: Coryanthes macrantha (izquierda), Masdevallia veitchiana (centro) y Dendrobium secundum (derecha).]

Diapositiva 4

Diapositiva 4: ¿Por qué hay que proteger las orquídeas?

Las orquídeas se ven muy amenazadas por la destrucción del hábitat y, en menor medida, por la recolección excesiva. Aunque la pérdida del hábitat afecta a todas las especies, la sobreexplotación es particularmente grave para aquellas especies de importancia comercial, y puede conducir a la extinción de una especie en la naturaleza a muy pocos años de su descubrimiento.

A través de la evolución, las orquídeas han desarrollado complejas estrategias de polinización fácilmente perturbadas por la destrucción del hábitat o por una recolección excesiva. Esto les hace buenos indicadores ambientales; entonces, si conseguimos salvar las orquídeas y mantener buenas poblaciones, probablemente estaremos salvando muchas otras especies a la vez. Al ser tan atractivas para el público, las orquídeas también constituyen el buque insignia de los esfuerzos encaminados a fomentar la conservación de hábitats críticos para la flora y fauna silvestres.

[Nota al ponente: Esta diapositiva muestra Phragmipedium besseae.]

Introducción a las zapatillas de Venus

Diapositiva 6: ¿Qué son las zapatillas de Venus?

Las zapatillas de Venus se distinguen fácilmente de otrs orquídeas por sus flores en forma de zapatilla o zapato, que han inspirado diversos nombres vernáculos, como "Lady's Slipper" (zapatito de dama) en Europa, "Moccasin Flower" (flor de mocasín) en Norteamérica y Zapatilla o Chinela en Latinoamérica. Los géneros más importantes de zapatillas de Venus son *Cypripedium*, *Paphiopedilum* y *Phragmipedium*. Todos estos géneros son solicitados en el comercio internacional.

El género *Cypripedium* incluye unas 50 especies cuya distribución atraviesa las regiones templadas septentrionales. El género *Paphiopedilum* se compone de aproximadamente 80 especies originarias del sudeste asiático, y el género *Phragmipedium* contiene alrededor de 20 especies cuya distribución se limita a América Central y del Sur. Sus flores llamativas y el número relativamente reducido de especies han hecho que estas plantas sean muy atractivas para viveristas y coleccionistas de orquídeas.

[Nota al ponente: La diapositiva muestra Phragmipedium var. dalessandroi (izquierda), Cypripedium parviflorum (centro) y Paphiopedilum venustum (derecha).]

Diapositiva 7: Zapatillas de Venus en CITES

Los géneros *Paphiopedilum* y *Phragmipedium* están incluidos en el Apéndice I de CITES. La inclusión en el Apéndice I en efecto prohíbe el comercio de plantas recolectadas en el medio silvestre, pero autoriza el comercio de plantas reproducidas artificialmente, siempre que vayan acompañadas del correspondiente permiso. Los dos géneros están incluidos en el Apéndice I de CITES como *Paphiopedilum* spp. y *Phragmipedium* spp. Esta "inclusión genérica" significa que las especies nuevas de estos géneros automáticamente entran en el Apéndice I de CITES en cuanto se describan, lo que garantiza el control inmediato sobre unas plantas recién descubiertas que son vulnerables al comercio insostenible.

El género *Cypripedium* está incluido en el Apéndice II de CITES, lo que significa que está autorizado el comercio de especímenes reproducidos artificialmente y de origen silvestre, cuando se hayan obtenido los permisos correctos. Sin embargo, la mayoría de los países (o Estados, en términos de CITES) del área de distribución han prohibido la exportación de *Cypripedium* recolectadas en la naturaleza. Además, en la normativa de la Unión Europea sobre comercio de especies silvestres que establece la aplicación de CITES para los 25 países miembros, *Cypripedium calceolus* recibe el mismo tratamiento que las especies incluidas en el Apéndice I de la Convención. Con estas medidas, resulta difícil encontrar plantas de *Cypripedium* recolectadas legalmente del medio silvestre en el comercio internacional, y si se encuentran, suelen ser ejemplares de las especies norteamericanas más comunes, procedentes de actividades de recolección controlada o de rescate.

[Nota al ponente: Para comprobar la lista más actualizada de los Apéndices CITES consulte la página web de CITES www.cites.org. Para informarse de los detalles de los controles más estrictos aplicados por la Unión Europea, visite la página web www.eu-wildlifetrade.org. Esta diapositiva muestra Phragmipedium longifolium (izquierda), Cypripedium himalaicum (centro) y Paphiopedilum henryanum (derecha).]

Diapositiva 8

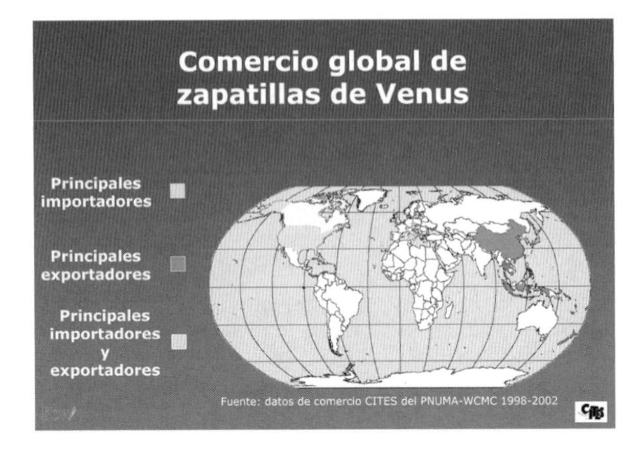

Diapositiva 8: Comercio global de zapatillas de Venus

Las zapatillas de Venus, sobre todo el género *Paphiopedilum*, están entre los cinco géneros de orquídeas más importantes para la horticultura. El volumen de comercio de estas plantas es muy elevado, y se presentan principalmente como ejemplares vivos de especies e híbridos cultivados por el hombre.

Entre 1998 y 2002 los datos comerciales CITES muestran más de 660.000 zapatillas de Venus en el comercio internacional. Casi todo este material objeto de comercio era reproducido artificialmente. Los registros de comercio de plantas silvestres corresponden a especímenes de *Cypripedium*.

Entre 1998 y 2002 los mayores exportadores de zapatillas de Venus reproducidas artificialmente, según estos registros, fueron Taiwán (Provincia de China), Indonesia y China, con más de la mitad (54%) de todas las exportaciones internacionales. Otros exportadores importantes durante el mismo periodo fueron la República de Corea, Países Bajos, Tailandia, Estados Unidos de América, Nueva Zelanda, Japón y Bélgica; todos ellos con más de 10.000 plantas exportadas.

El mayor importador de zapatillas de Venus reproducidas artificialmente entre 1998 y 2002 fue Japón, con más de la mitad (56%) de todas las importaciones. Otros importadores importantes durante el mismo periodo fueron Canadá, la República de Corea, Estados Unidos de América, Alemania, Italia y Suiza; todos ellos con más de 10.000 importaciones.

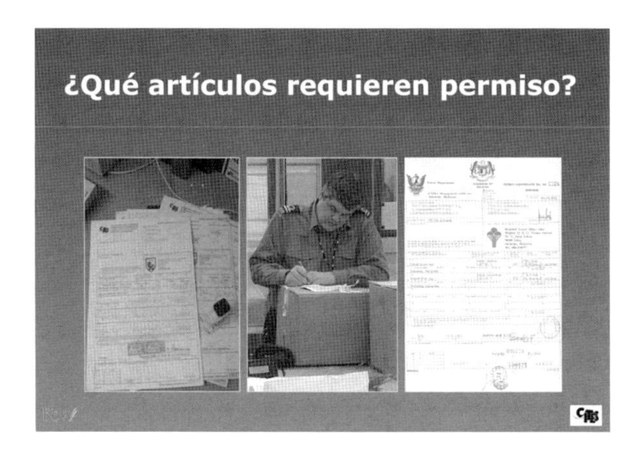

Diapositiva 9: ¿Qué artículos requieren permiso?

La respuesta simple es ¡todos los que no estén exentos! Los controles de la CITES se aplican a las plantas "vivas o muertas" y a sus "partes y derivados fácilmente identificables".

Esto quiere decir que no son sólo las plantas enteras las que están sujetas al control de la Convención, sino también las partes de las plantas, como semillas, estacas, y hojas. Asimismo, el control puede abarcar los productos elaborados a partir de plantas. Si el nombre de una planta o animal incluido en los Apéndices CITES figura escrito en el embalaje de un producto, entonces se considera que dicho producto contiene la especie y por tanto está sujeto a las disposiciones del Convenio.

Para las plantas incluidas en el Apéndice I de CITES, se controlan tanto la planta entera como todas sus partes y derivados, vivos o muertos. Tan sólo existe una exención: "los cultivos de plántulas o de tejidos obtenidos *in vitro*, en medios sólidos o líquidos, que se transportan en envases estériles no están sujetos a las disposiciones de la Convención". No es necesario que este material vaya en frascos o botellas tradicionales para cumplir los criterios de la exención; sólo tienen que ser envases estériles. Ahora bien, el material debe ser de origen legal para ser exento. En la próxima diapositiva, veremos los detalles relativos al control de cultivos de tejidos.

Para las plantas incluidas en el Apéndice II de CITES, se controla el ejemplar "vivo o muerto" y las partes y derivados fácilmente identificables especificados en los Apéndices. En el caso de las *Cypripedium*, las únicas partes y derivados exentos del control del Convenio son: a) las semillas y el polen (inclusive las polinias); b) los cultivos de plántulas o de tejidos obtenidos *in vitro*, en medios sólidos o líquidos, que se transportan en envases estériles; y c) las flores cortadas de ejemplares reproducidos artificialmente.

[Nota al ponente: La diapositiva muestra permisos CITES y un aduanero comprobando documentos.]

Diapositiva 10

Diapositiva 10: Cultivos de tejidos

Después de incluirse *Paphiopedilum* y *Phragmipedium* en el Apéndice I de CITES, la Conferencia de las Partes (CdP) aprobó eximir los cultivos de tejidos de estos géneros de las disposiciones de la Convención. Fue una exención insólita – exonerar del control a una parte o derivado "fácilmente identificable" de un espécimen del Apéndice I. Fueron muchas las Partes que se opusieron a la exención, pero se aprobó por el voto de la mayoría. La exención tiene la finalidad de alentar la reproducción artificial de estas plantas tan solicitadas, para reducir la presión que pesa sobre sus poblaciones naturales.

En su momento, se creía que esta reproducción nunca podría ser perjudicial para las poblaciones silvestres, pero últimamente se ha demostrado que no es así. La recolección ilegal e insostenible de especies endémicas de Vietnam, en particular, es causa de preocupación. Algunos países han expresado inquietudes respecto de la exención de cultivos de plántulas y tejidos en frascos, que pueden haber aprovechado comerciantes poco escrupulosos para "blanquear" el comercio de material cuyo plantel parental se ha recolectado de manera ilícita del medio silvestre.

El plantel parental empleado para producir cultivos de tejidos debe obtenerse de forma legal, según la normativa vigente para la especie en cuestión en su país de origen. Si este plantel es ilegal, los cultivos de tejidos que de él se deriven no cumplen los criterios para la exención del control de CITES, y entonces pueden ser confiscados por las autoridades encargadas del cumplimiento del Convenio.

[Nota al ponente: La diapositiva muestra cultivos de plántulas en envases estériles: frascos, bolsas y botellas.]

Zapatillas de Venus
en CITES

Diapositiva 12

Diapositiva 12: ***Cypripedium***

El género Cypripedium consta de unas 50 especies que se encuentran en las zonas templadas septentrionales de Asia, Europa y Norteamérica, extendiéndose hacia el sur hasta Honduras, Guatemala y la región meridional de China. Se dan en una amplia gama de hábitats, desde zonas boscosas con coníferas o bosques mixtos con árboles caducifolios, hasta pantanos y pastizales. Son terrestres, con hojas que en la mayoría de las especies salen nuevas de la base cada año. Las flores son como zapatillas, y sus colores cubren toda la gama desde el verde, pasando por el blanco y el amarillo, hasta el rojo y púrpura intensa. A los jardineros de climas templados les gustan las *Cypripedium*, pues todas tienen cierto grado de resistencia al frío y pueden cultivarse en exteriores al menos parte de la temporada.

[Nota al ponente: La diapositiva muestra Cypripedium parviflorum (izquierda), Cypripedium guttatum (centro superior), Cypripedium lichiangense (centro inferior) y Cypripedium smithii (derecha).]

Diapositiva 13: Características de las *Cypripedium* – la flor

Todas las flores del género *Cypripedium* tienen forma de zapatilla. Las flores pueden presentarse solas o múltiples, de color verde, o blancas con manchas púrpuras, o doradas, o de un tono pardo-malva. Las flores no se caen, sino que permanecen en el fruto.

[Nota al ponente: La diapositiva muestra, de izquierda a derecha en la fila superior Cypripedium acaule, Cypripedium palangshanense, Cypripedium japonicum, Cypripedium wardii, y Cypripedium lichiangense; de izquierda a derecha en la fila inferior Cypripedium arietinum, Cypripedium reginae, Cypripedium irapeanum, Cypripedium x froschii y Cypripedium calceolus.]

Diapositiva 14

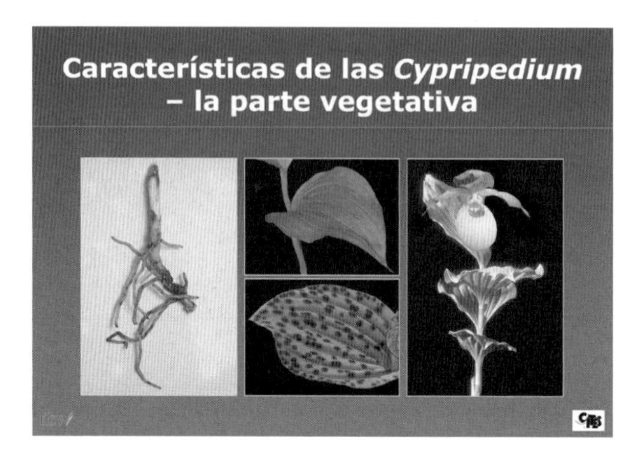

Diapositiva 14: Características de las *Cypripedium* – la parte vegetativa

- Todas las *Cypripedium* tienen un distintivo tallo subterráneo modificado, semejante a una raíz, llamado rizoma.

- En la mayoría de las especies, este rizoma es corto y rara vez se ramifica, sino que produce lo que parece un rosario de puntos de crecimiento anuales.

- Las hojas perecen anualmente, dejando el rizoma para sobrevivir el periodo latente anual. En primavera las nuevas hojas nacen de las yemas del rizoma.

- Las hojas normalmente son de forma oval.

- Las hojas están plegadas por su eje largo, y tienen nervaduras pronunciadas.

- Las hojas a menudo tienen vello, sobre todo en las nervaduras y en los márgenes.

[Nota al ponente: La diapositiva muestra un rizoma con la yema y las raíces (izquierda), hojas de colores variados (centro superior e inferior), tallo, hojas y flor de Cypripedium fasciolatum (derecha).]

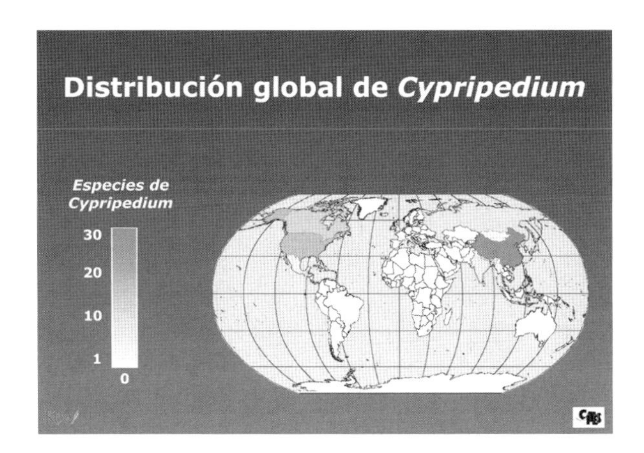

Diapositiva 15: Distribución global de *Cypripedium*

Aunque el género Cypripedium se encuentra en todas las zonas templadas septentrionales del mundo, en China es donde se da el mayor número de especies, con diferencia, de las más solicitadas (debido a su difícil acceso en el pasado). La mayor parte de las especies que ahora se cultivan ampliamente proviene de Norteamérica; también es común el cultivo de la especie europea *Cypripedium calceolus*, y la japonesa *Cypripedium formosanum*.

Diapositiva 16

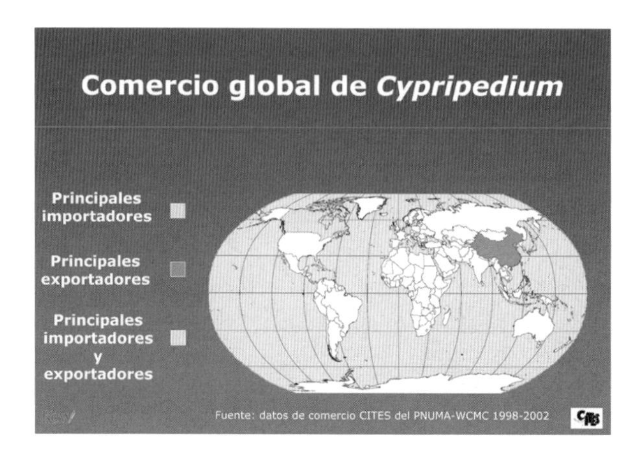

Diapositiva 16: Comercio global de *Cypripedium*

La inmensa mayoría del comercio CITES de taxa de *Cypripedium* registrado en los años 1998 - 2002 se dio en plantas vivas. Según informes, prácticamente siempre se trataba de especímenes reproducidos artificialmente (>98 %).

La mayor parte del material reproducido artificialmente que figuraba en estos informes provenía de Taiwán (Provincia de China), China y la República de Corea. Estos tres exportadores fueron los proveedores del 93 por ciento de los ejemplares reproducidos artificialmente registrados entre 1998 y 2002.

El principal importador de *Cypripedium* en el mismo periodo fue Japón, con el 78 por ciento de todas las importaciones registradas en los datos CITES. Japón, la República de Corea, Canadá y Alemania acapararon el 95 por ciento de todas las importaciones.

El comercio de plantas recolectadas en la naturaleza entre 1998 y 2002, según los informes, no llegó al 2 por ciento del total. Los principales proveedores de este material fueron los Estados Unidos de América, la Federación Rusa y China, siendo Alemania, Japón y el Reino Unido los mayores importadores de especímenes de origen silvestre.

[Nota al ponente: Los datos comerciales CITES se pueden descargar de la Base de datos de comercio CITES del PNUMA-WCMC, accesible en Internet a través de la página web de la Secretaría CITES: www.cites.org]

Diapositiva 17: ***Cypripedium lichiangense* y *C. palangshanense***

Originaria del noreste de Myanmar y del noroeste de Yunnan y suroeste de Sichuan, en China, *Cypripedium lichiangense* fue descrita en 1994. Es una de las especies más buscadas, con la flor moteada como el huevo de un pájaro que descansa sobre un par de hojas. Esta especie es particularmente difícil de cultivar y reproducir. Las plantas cultivadas fuera de su hábitat natural no suelen sobrevivir más de 3-4 años.

Aunque *Cypripedium palangshanense* fue descrita en 1936, esta especie, con pequeñas flores púrpuras, permaneció prácticamente desconocida para el mundo occidental hasta que se redescubrió en 1998. Es endémica de un área de distribución muy pequeña, entre el noroeste y el este de Sichuan, en China. Tiene un rizoma rastrero muy delgado.

Ambas especies han llegado a entrar en el comercio.

[Nota al ponente: La diapositiva muestra Cypripedium lichiangense (izquierda) y Cypripedium palangshanense (superior e inferior derecha).]

Diapositiva 18: *Phragmipedium*

Phragmipedium es un género pequeño que contiene unas 20 especies. Su área de distribución abarca desde el sur de México y Guatemala, a través de América Central y del Sur hasta la parte meridional de Bolivia y Brasil. Pueden crecer en el suelo, o apoyarse en superficies rocosas o en árboles. Típicamente se encuentran estas plantas alrededor de cascadas y en otras zonas húmedas.

El rizoma (tallo modificado semejante a una raíz) es rastrero. Las hojas son coriáceas, y en sección transversal tienen forma de V. Estas orquídeas son menos solicitadas en el comercio que sus rivales del sudeste asiático, las *Paphiopedilum*, pero el descubrimiento de *Phragmipedium besseae* en los años 1980, con sus novedosas flores rojas, provocó un marcado aumento del comercio y del interés de los coleccionistas por este género. Más recientemente, en 2002, el descubrimiento de *Phragmipedium kovachii* en Perú, con sus grandes flores púrpuras, suscitó aún más interés en este grupo.

[Nota al ponente: La diapositiva muestra Phragmipedium besseae (izquierda y centro superior) y Phragmipedium longifolium (derecha y centro inferior).]

Diapositiva 19: Características de las *Phragmipedium* – la flor

Antes de florecer las yemas, los márgenes de las hojas modificadas (los sépalos, que parecen pétalos) que rodean la flor se tocan. En los demás grupos los sépalos se imbrican. Además, el margen del labio de la flor en forma de zapatilla se dobla hacia dentro. A diferencia de las *Cypripedium* la flor es caduca, desprendiéndose de la punta del fruto al marchitarse.

[Nota al ponente: La diapositiva muestra Phragmipedium besseae var. dalessandroi (izquierda), Phragmipedium kovachii (centro izquierda superior), Phragmipedium longifolium (centro derecha superior), Phragmipedium lindenii (derecha), Phragmipedium schlimii (centro izquierda inferior), Phragmipedium wallisii (centro derecha inferior).]

Diapositiva 20: Características de las *Phragmipedium* – la parte vegetativa

Las *Phragmipedium* tienen las hojas de oblongas a lineares, normalmente de color verde apagado o con brillo, y sin marcas. Son aplanadas, coriáceas, se presentan en dos hileras o filas en caras opuestas del tallo, y la costilla media se ve pronunciada pero no las nervaduras. Las hojas persisten dos o más años; no se caen cada año. Típicamente son largas, liguladas, y acuminadas. Los tallos son característicamente cortos. Todas las especies tienen un rizoma corto o rastrero que carece de los nudos o puntos de crecimiento anuales que se observan en las *Cypripedium*.

[Nota al ponente: La diapositiva muestra Phragmipedium besseae (izquierda), Phragmipedium lindleyanum (centro superior), Phragmipedium longifolium (centro inferior) y las puntas de las hojas de una especie de Phragmipedium (derecha).]

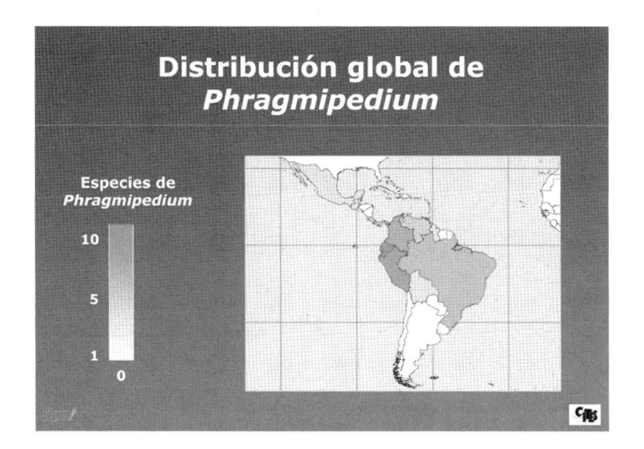

Diapositiva 21: Distribución global de *Phragmipedium*

El género se encuentra en América Central y del Sur. Su área de distribución se extiende desde el sur de México y Guatemala hasta el sur de Bolivia y Brasil.

Diapositiva 22

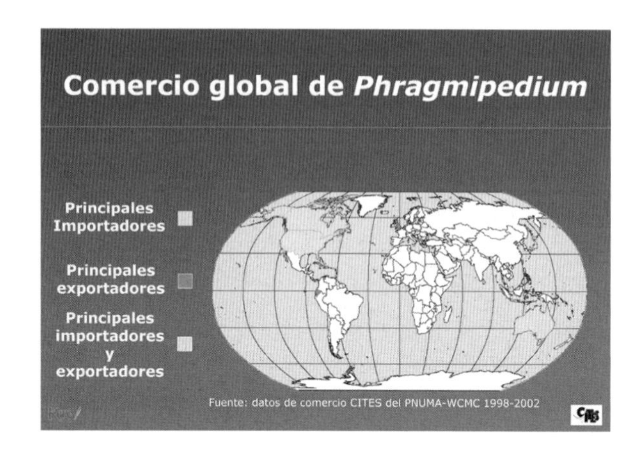

Diapositiva 22: Comercio global de *Phragmipedium*

El comercio registrado de taxa de *Phragmipedium* durante los años 1998 - 2002 no llegó al 2 por ciento de todo el comercio internacional de zapatillas de Venus. Todos los registros de comercio CITES de taxa de *Phragmipedium* en ese periodo se referían a material reproducido artificialmente.

Ecuador, el Reino Unido, Taiwán (Provincia de China) y los Estados Unidos de América fueron los principales exportadores de plantas reproducidas artificialmente durante esos años, acaparando el 84 por ciento del total. A Japón, Estados Unidos de América, Canadá y Australia llegó el 77 por ciento de todas las *Phragmipedium* importadas en ese periodo. Los datos comerciales entre 1998 y 2002 también indican que muchos países recibieron sus ejemplares reexportados de los Estados Unidos de América.

[Nota al ponente: Los datos comerciales CITES se pueden descargar de la Base de datos de comercio CITES del PNUMA-WCMC, accesible en Internet a través de la página web de la Secretaría CITES: www.cites.org.]

Diapositiva 23: ***Phragmipedium kovachii***

El descubrimiento y la posterior importación en EE. UU. de *Phragmipedium kovachii* ha generado una publicidad considerable. Se descubrió en el norte de Perú y fue descrito en 2002. En interés de los medios de comunicación por esta especie ocultó el hecho de que se creía que su extracción por parte de coleccionistas la había llevado a la extinción. Sin embargo, de los 5 sitios conocidos, al menos uno aún existía en 2004, con varios centenares de plantas. Según informes, el Gobierno de Perú ha aprobado una recolección muy limitada de especímenes del medio silvestre para permitir la reproducción controlada a partir de semillas. Se ha comunicado que Perú permite la exportación de especies e híbridos selectos de al menos un vivero autorizado. El éxito de la conservación de esta especie tan excepcional y solicitada sólo puede verse complementado por una reproducción masiva de plántulas, tan pronto como sea posible, para menoscabar el valor de las plantas ilegales de origen silvestre.

[Nota al ponente: Esta diapositiva muestra tres imágenes de Phragmipedium kovachii.]

Diapositiva 24

Diapositiva 24: *Paphiopedilum*

Paphiopedilum es el género más grande de zapatillas de Venus, con unas 80 especies naturales de los trópicos asiáticos, desde el sur de la India hasta Nueva Guinea y Filipinas. Se pueden encontrar en el suelo, en desfiladeros, en superficies rocosas, y agarradas a árboles u otra vegetación, pero la mayoría son terrestres. Crecen en hojarasca o en grietas de rocas que contienen materia orgánica. Están presentes en hábitats muy variados, desde las ramas de los grandes árboles de las pluviselvas de Tailandia hasta los suelos escarpados y sinuosos del Monte Kinabalu.

Todas tienen la característica flor que parece una zapatilla; algunas con flores exageradamente infladas, como la *Paphiopedilum micranthum*, (que en inglés se llama "orquídea chicle"), en la imagen derecha de esta diapositiva. Constituyen el grupo más popular de zapatillas de Venus para los coleccionistas y viveristas. Se encuentran entre los 5 géneros de orquídeas más importantes en el negocio de la horticultura, produciéndose numerosos híbridos artificiales cada año. No obstante, ha existido un nivel significativo de recolección y comercio ilícitos de especies de *Paphiopedilum*.

[Nota al ponente: Esta diapositiva muestra Paphiopedilum exul (izquierda), Paphiopedilum rothschildianum (centro) y Paphiopedilum micranthum (derecha).]

Diapositiva 25: Características de las *Paphiopedilum* – la flor

Antes de florecer las yemas, las hojas modificadas (los sépalos, que parecen pétalos) que rodean la flor se imbrican. Además, el margen del labio de la flor en forma de zapatilla no se dobla. A diferencia de *Cypripedium*, la flor es caduca y se cae de la punta del fruto al marchitarse. En muchas especies las flores tienen vello y protuberancias o callosidades como verrugas. Muchas especies producen una sola flor, pero algunas portan múltiples flores. La flor presenta una amplia gama de colores; verde, blanca, dorada pura, y púrpura. Algunas de las especies más buscadas tienen flores con largos pétalos colgantes que llegan al extremo de *Paphiopedilum sanderianum*, de más de un metro de largo.

[Nota al ponente: La diapositiva muestra Paphiopedilum malipoense (izquierda), Paphiopedilum haynaldianum (centro izquierda superior), Paphiopedilum bellatulum (centro derecha superior), Paphiopedilum druryi (centro izquierda inferior), Paphiopedilum liemianum (centro derecha inferior) y Paphiopedilum sanderianum (derecha).]

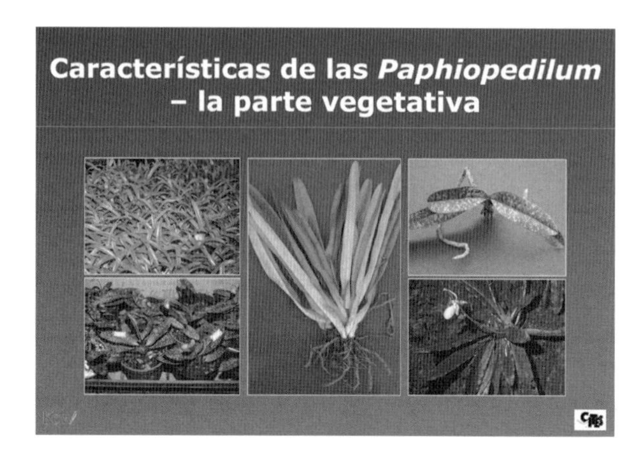

Diapositiva 26: Características de las *Paphiopedilum* – la parte vegetativa

Un rizoma está presente en todas las especies, pero normalmente es corta. En pocas especies, *P. bullenianum*, *P. armeniacum*, *P. micranthum* y *P. druryi*, el rizoma es rastrero y puede llegar a medir un metro de largo.

Las hojas de las *Paphiopedilum* son coriáceas, con la costilla media pronunciada, y en sección transversal tienen forma de V. Las hojas pueden ser cortas y liguladas o de oblongas a lineares, pero habitualmente son cortas, de menos de 20 centímetros de largo. El grupo de flores múltiples, que incluye especies como *Paphiopedilum sanderianum*, *P. rothschildianum* y *P. lowii*, constituye una excepción a esta regla. Se pueden clasificar varios tipos de color de hoja, lo que resulta muy útil para la identificación. Los colores de las hojas van desde el verde, mate o con brillo, hasta un tono de púrpura moteada.

[Nota al ponente: La diapositiva muestra plantas con hojas verdes coriáceas y liguladas (izquierda superior), hojas moteadas (izquierda inferior), hojas que superan los 20 cm de largo (centro), hojas moteadas (derecha superior) y hojas liguladas verdes con brillo en el hábitat (Paphiopedilum liemianum, derecha inferior).]

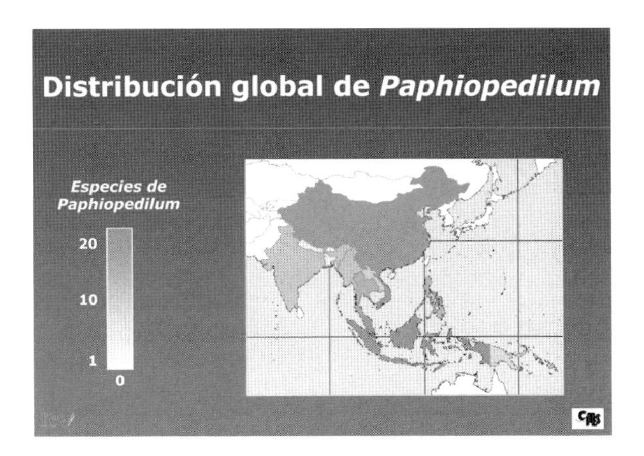

Diapositiva 27: Distribución global de *Paphiopedilum*

El género *Paphiopedilum* incluye unas 80 especies. Muchas son endémicas de áreas de distribución muy reducidas, pero el género entero se da en todo el sudeste asiático. Se encuentran desde la India hacia el este a través de China a las Filipinas, y hacia el sur por el archipiélago malayo hasta Nueva Guinea y las Islas Salomón. Muchas de las especies más buscadas son autóctonas de China y Vietnam.

Diapositiva 28

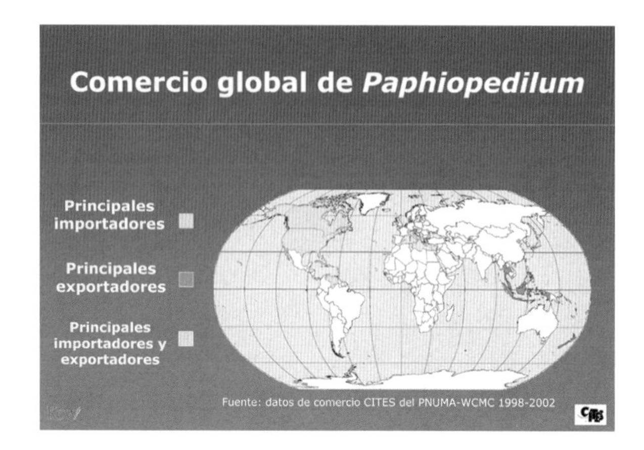

Diapositiva 28: Comercio global de *Paphiopedilum*

Países muy diversos comunicaron exportaciones de taxa de *Paphiopedilum* artificialmente reproducidas entre 1998 y 2002. Los mayores exportadores fueron Indonesia, Países Bajos, Tailandia y Nueva Zelanda, con el 79 por ciento de todas las exportaciones registradas. Además, Taiwán (Provincia de China), Estados Unidos de América, Japón y Bélgica exportaron más de 10.000 plantas cada uno, así que los 8 principales exportadores acapararon algo más del 90 por ciento del total. Con Japón, Malasia y Austria, esa cifra supera el 97 por ciento de todas las exportaciones.

Los dos principales importadores entre 1998 y 2002 fueron Japón y Estados Unidos de América, acaparando el 75 por ciento de todas las importaciones registradas entre los dos. Estos estados juntos con Canadá, Italia, Suiza, Hong Kong y Venezuela importaron el 93 por ciento del volumen total durante ese periodo.

[Nota al ponente: Los datos comerciales de CITES se pueden descargar de la Base de datos de comercio CITES del PNUMA-WCMC, accesible en Internet a través de la página web de la Secretaría CITES: www.cites.org]

Diapositiva 29: China y Vietnam

Entre los años 1980 y primeros 1990 se encontró un grupo nuevo de *Paphiopedilum* en China. No sólo supuso un apasionante descubrimiento nuevo, con especies diferentes de todas las que se habían conocido anteriormente, sino que también representaba toda una nueva línea en potencia para el cultivo de híbridos. Las especies recién descubiertas incluían *Paphiopedilum armeniacum, P. emersonii, P. micranthum* y *P. malipoense.* El resultado fue un volumen importante de comercio ilícito. Ahora su cultivo está bastante bien establecido, lo que ha tenido el efecto de reducir la necesidad de extraerlas de la naturaleza, aunque siempre se siguen buscando formas nuevas, de colores insólitos, como por ejemplo orquídeas albinas. De todos modos, la presión a causa de la recolección de estas plantas para el mercado internacional las ha hecho vulnerables de extinción en China.

Hacia finales de los 1990 y primeros de los 2000 se describieron algunas especies nuevas y poco comunes de zapatillas de Venus en Vietnam, de las cuales las más notables son *Paphiopedilum vietnamense* y *P. hangianum.* Ambas presentan formas novedosas, por lo que son muy solicitadas. *Paphiopedilum vietnamense* fue descrita por primera vez en 1999. En 2001 se lanzó una expedición para explorar la única localidad conocida, encontrándose sólo un manojo de plántulas. Debido al área de distribución tan restringida, y a su nivel de explotación, esta especie ya se considera En Peligro Crítico.

[Nota al ponente: La diapositiva muestra Paphiopedilum armeniacum (izquierda), Paphiopedilum emersonii (centro superior), Paphiopedilum malipoense (centro inferior) y Paphiopedilum micranthum (derecha).]

Diapositiva 30

Diapositiva 30: Zapatillas de Venus en CITES: resumen

En esta sección hemos presentado:

- los tres géneros de zapatillas de Venus incluidos en CITES, *Cypripedium*, *Phragmipedium* y *Paphiopedilum;*

- sus características, distribución global, y comercio.

[Nota al ponente: La diapositiva muestra Phragmipedium wallisii (izquierda), Cypripedium flavum (centro) y Paphiopedilum helenae (derecha).]

Aplicación de CITES para zapatillas de Venus

Diapositiva 32

Diapositiva 32: Cumplimiento

Los controles de CITES se hacen cumplir a distintos niveles. Dentro de un país exportador, se efectúan inspecciones de viveros, comerciantes, mercados y, con menos frecuencia pero más importancia, de las plantas en el momento de la exportación. Las inspecciones también pueden ocurrir en el momento de la importación y después de la importación en los principales países comerciantes. Además, los organismos encargados del cumplimiento vigilan ferias comerciales, anuncios en la prensa comercial y en Internet.

Pocos países cuentan con equipos especialmente entrenados en la identificación de especímenes CITES, ni animales ni plantas. Habitualmente son los aduaneros en general, o agentes con formación en controles fitosanitarios, los encargados del cumplimiento del Convenio. Cuando son los aduaneros en general, los procedimientos se centran en la documentación y no en las plantas. Comprueban que los permisos estén correctamente rellenados, sellados, y emitidos por las autoridades correctas. También verifican otros documentos y facturas para ver si el material CITES que figura en la documentación acompañante corresponde con los permisos CITES.

Cuando son estos aduaneros los que inspeccionan las plantas CITES, es imprescindible que tengan contacto con un centro de expertos en la identificación y conservación de plantas. Este tipo de centro debería ser la Autoridad Científica nacional, pero en algunos casos dicha Autoridad puede ser una comisión o departamento gubernamental con más conocimiento de animales. En este caso, los organismos encargados del cumplimiento tendrían que establecer relaciones con un jardín botánico o herbario nacional o local. Este tipo de contacto es de vital importancia.

Los aduaneros necesitarán cierta formación básica sobre las plantas y sus partes y derivados amparados por CITES, y les hará falta ayuda para atajar el comercio perjudicial. Lo más importante es que los aduaneros deben tener contacto con expertos capaces de identificar las plantas CITES, que puedan dar consejos y tener acceso a instalaciones para guardar material decomisado o confiscado. Estos científicos también pueden actuar como testigos expertos, imprescindibles en los tribunales cuando las infracciones de los controles son procesadas.

Diapositiva 32

[Nota al ponente: Para identificar a la persona competente dentro de la plantilla de la Secretaría CITES a quien contactar sobre cuestiones de cumplimiento, compruebe la lista de personal en la página web de CITES: www.cites.org.]

Diapositiva 33

Diapositiva 33: Cumplimiento – comprobaciones

Documentos – Comprobar la autenticidad de los permisos CITES (firmas, sellos), los nombres de las plantas y el número de especímenes, y asegurar que los nombres de las plantas y el número de especímenes que constan en los permisos coincidan con los que figuran en los albaranes. También hay que comprobar el origen de las plantas. ¿Se declaran reproducidas artificialmente o de origen silvestre? ¿La planta es una especie que se acaba de describir? Haga Ud. uso de las bases de datos y listas de referencia recomendadas en la Sección "Referencias y recursos". ¿Se trata de plántulas en frascos o cultivos de tejidos, supuestamente exentos de las disposiciones de la Convención? Si es así, y son especies recién descubiertas, es posible que Ud. tenga que pedir a su Autoridad Administrativa que confirme el origen legal del plantel parental.

País de origen – Siempre se debe comprobar el país de origen que consta en los permisos. ¿Se están exportando orquídeas de un país donde las plantas se den en la naturaleza? Si es así, es más probable que los ejemplares en cuestión sean de origen silvestre. Los países pueden expresar su preocupación sobre la exportación ilegal de sus zapatillas de Venus recolectadas en el medio silvestre, y pedir la ayuda de otras Partes y Estados no Partes en la Convención para controlar dicho comercio. Normalmente, este tipo de petición se publica como una Notificación a las Partes en CITES (se puede encontrar en la página web de CITES: www.cites.org). Vietnam, por ejemplo, es un país que ha manifestado inquietudes sobre el comercio internacional ilícito de especies autóctonas de *Paphiopedilum*.

Embalaje – Los viveros suelen envolver y empaquetar sus plantas con cuidado, para evitar dañarlas. Luego se transportan en cajas marcadas con el nombre del vivero y etiquetas impresas. Las partidas de plantas ilegalmente recolectadas pueden ir empaquetadas con materiales locales y el embalaje puede ser de peor calidad, con etiquetas escritas a mano (a veces con datos de la recolección). Puede que los especímenes no estén identificados a nivel de especie, para encubrir la posible extracción de especies nuevas sin nombre.

Envíos de plantas – Las colecciones ilícitas de plantas habitualmente se componen de pequeñas muestras de ejemplares de distintos tamaños y edades y de formas desiguales. Estas plantas pueden estar dañadas (con las raíces rotas o partidas), y pueden llevar restos de suelo y maleza u otro material vegetal autóctono entre los tallos y raíces. Los especímenes reproducidos artificialmente son uniformes, y los envíos vienen limpios, sin parásitos ni restos de plagas, enfermedades, suelo, malas hierbas, o plantas autóctonas.

Rutas comerciales y contrabando – Las colecciones ilegales de especies raras o nuevas se pueden mandar por servicios postales o de mensajería, o se pueden llevar en el equipaje de mano para evitar ser detectadas. También pueden repartirse entre varios paquetes y enviarse por separado, para garantizar un alto índice de supervivencia, además de que al menos algunas de las plantas escapen sin ser descubiertas.

Diapositiva 34

Diapositiva 34: Reproducidas artificialmente o de origen silvestre, ¿cómo distinguir?

No es tan sencillo distinguir entre plantas reproducidas artificialmente y las de origen silvestre. Sin embargo, existen ciertas características que pueden servir para determinar la diferencia.

Las plantas recolectadas en el medio silvestre llevan las señales de haber sido arrancadas de su hábitat natural. En cambio, las plantas cultivadas en viveros llevan las señales de un ambiente artificial, bien cuidado. Se ven limpias, uniformes, y empaquetadas con arreglo a normas exigentes. A veces se cultivan orquídeas en el exterior o en umbráculos, y entonces pueden detectarse algunos signos parecidos a los de las plantas de origen silvestre. Por eso es importante llamar a un experto que compruebe la situación de un envío de plantas, si se sospecha que pueden ser de origen silvestre y no reproducidas artificialmente.

El Manual de identificación CITES, volumen 1, flora, disponible de la Secretaría CITES, incluye detalles sobre cómo distinguir entre plantas reproducidas artificialmente y las de origen silvestre, para los principales grupos amparados por la Convención. Pero recuerde, ¡que su opinión siempre sea verificada por un experto!

Diapositiva 35: Orquídeas recolectadas en el medio silvestre

Las raíces de las plantas recolectadas en la naturaleza suelen estar muertas, o se han roto o recortado de manera tosca en un esfuerzo de limpiar la planta después de su recolección. Pueden crecer nuevas raíces a partir del sistema radicular viejo y dañado. También es posible encontrar todavía restos del sustrato natural adheridos a las raíces de plantas de origen silvestre. Además, hay que comprobar el orden de los restos adheridos a las raíces. Por ejemplo, las raíces pueden presentar alguna materia orgánica directamente adherida, luego puede haber algo de musgo esfagno utilizado en el transporte, y finalmente es posible encontrar algún tipo de abono hortícola, como corteza o lana de roca. Pero recuerde, siempre se debe hacer la evaluación con cautela.

Las hojas de las plantas recolectadas en el medio silvestre demuestran las marcas de su hábitat natural, los daños sufridos por la extracción, y frecuentemente se ve el contraste con nuevos brotes que hayan salido después de la recolección. Las hojas basales suelen estar muertas o dañadas. Pueden estar picadas las hojas por desecación y llevar las huellas de insectos horadadores. Las plantas recién arrancadas también pueden tener matas de musgo, líquenes o hepáticas. Este tipo de vegetación normalmente no sobreviviría en las condiciones controladas de un vivero de orquídeas. A medida que vayan creciendo las plantas silvestres después de entrar en un vivero, brotan nuevas hojas de aspecto limpio y fresco, destacándose el contraste con las hojas viejas, "silvestres". Las hojas viejas, "silvestres", se pueden haber cortado deliberadamente para dejar sólo las pocas hojas nuevas que salieron durante el tiempo que estuvo la planta en el vivero.

El Manual de Identificación CITES, Volumen I, Flora incluye información detallada sobre cómo distinguir entre orquídeas reproducidas artificialmente y las de origen silvestre, pero siempre es importante que un experto dé una segunda opinión para confirmar que las plantas sean de origen silvestre. Las plantas cultivadas al aire libre, en condiciones deficientes o en umbráculos a veces presentan algunos de los rasgos típicos de plantas extraídas del medio silvestre.

[Nota al ponente: El Comité de Flora de CITES ha producido una serie de guías regionales con los nombres de personas expertas en CITES para contactar en los distintos países (véase la página web de CITES para más detalles). Posiblemente la guía le sirva a Ud. para establecer contacto con un experto relevante. Las

características de orquídeas reproducidas artificialmente y de origen silvestre se explican en el Manual de Identificación CITES, Volumen 1, flora. La Secretaría CITES manda copias a todas las Autoridades CITES. Si su Autoridad no tiene una copia actualizada del Manual, póngase en contacto con la Secretaría CITES.]

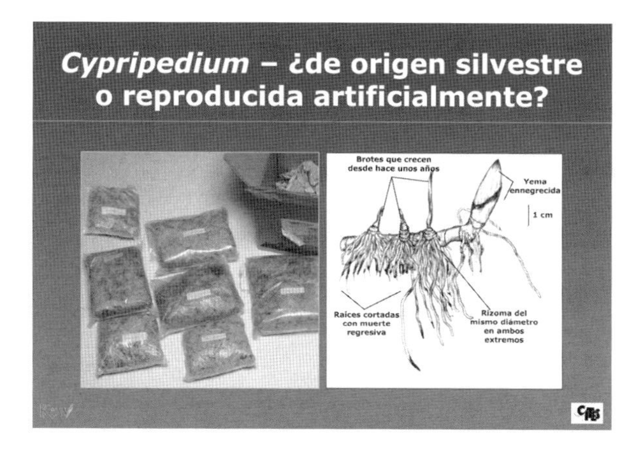

Diapositiva 36: *Cypripedium* – ¿de origen silvestre o reproducida artificialmente?

No es tan sencillo distinguir entre plantas reproducidas artificialmente y las de origen silvestre. Sin embargo, existen ciertas características que pueden servir para determinar la diferencia.

Todas las *Cypripedium* tienen un tallo subterráneo modificado, parecido a una raíz, llamado rizoma. El rizoma produce lo que parece un rosario de puntos de crecimiento anuales. En la mayoría de las especies, el rizoma es rastrero, corto pero robusto, y no suele ramificarse. En algunas especies, como *Cypripedium guttatum* y *C. margaritaceum*, el rizoma es alargado y los puntos de crecimiento anuales se presentan cada pocos centímetros. El rizoma sobrevive el periodo latente y la yema nueva ocupa una posición terminal. Las verdaderas raíces son fibrosas y emergen desde detrás del brote.

Las *Cypripedium* normalmente se comercian en estado latente, en primavera u otoño, como rizomas con yemas y con raíces fibrosas. Puede parecer que una partida grande no sea más que un saco de raíces, que no guarde la más mínima relación con lo que el gran público considere la orquídea típica. Las partidas comerciales de alta calidad a menudo se envuelven en musgo esfagno como se ve en la fotografía de la izquierda.

El rizoma es extremadamente útil, ya que puede servir como un indicio de la edad de la planta. Las cicatrices redondas del crecimiento del año anterior permanecen, y así es posible determinar la edad mínima de la planta. Un diámetro constante del rizoma indica una planta madura (normalmente al menos de cinco años de edad). En las plantas inmaduras, el rizoma aumenta paulatinamente de diámetro hasta alcanzar su tamaño óptimo.

Si al inspeccionar un envío, Ud. sospecha que la declaración sea errónea y que no sean plantas reproducidas artificialmente, debe contactar con un experto para confirmar su opinión.

Diapositiva 37

Diapositiva 37: *Cypripedium* – posible comercio ilícito

Es más probable que el comercio ilícito de orquídeas del género *Cypripedium* se dé con las especies descritas más recientemente y con aquellas que sean difíciles de cultivar. Pueden entrar en el comercio declaradas como plantas reproducidas artificialmente. Al verse casi siempre en el mercado internacional como rizomas, será necesaria la opinión de un experto para determinar si cumplen la definición CITES de "reproducidas artificialmente" o si son de origen silvestre. Para mantenerse al tanto de las especies de más reciente descripción, sugerimos que compruebe Ud. las bases de datos que hemos señalado en la Sección de "Referencias y recursos". Vea la fecha de la descripción científica original de la planta, y si es reciente, mayor es la probabilidad de que se haya extraído de la naturaleza.

[Nota al ponente: La diapositiva muestra bolsas de rizomas de Cypripedium de origen silvestre (izquierda) y C. x froschii (derecha).]

Paphiopedilum y Phragmipedium ¿de origen silvestre o reproducidas artificialmente?

	Origen silvestre	Reproducción artificial
Aspecto general	• Forma y tamaño irregulares • Dañadas y marcadas	• Forma y tamaño uniformes • Limpias y sanas
Raíces	• Dañadas o cortadas • Empiezan a crecer de nuevo	• Limpias y sanas • Pueden tener forma de maceta
Hojas	• Hojas inferiores dañadas o cortadas • Daños por insectos/horadadores • Picadas por desecación • Presencia de líquenes	• Limpias, sin daños • No recortadas • Prácticamente sin daños por insectos
Suelo	• Suelo o sustrato del hábitat adherido	• Sólo abono hortícola

Diapositiva 38: *Paphiopedilum* y *Phragmipedium* – ¿de origen silvestre o reproducidas artificialmente?

Esta diapositiva resume las características clave de las *Paphiopedilum* y *Phragmipedium* reproducidas artificialmente y de las de origen silvestre. Las orquídeas no suelen entrar en el comercio internacional en estado de floración, así que la primera identificación, además de la decisión si son de origen silvestre, tienen que basarse en un examen del material vegetativo. La preocupación inicial no debe ser tanto el nombre de la especie de orquídea, sino determinar si la planta fue extraída de la naturaleza. Si presenta bastantes de los rasgos señalados en esta diapositiva, y Ud. cree que puede ser de origen silvestre, debe llamar a un experto para confirmarlo

[Nota al ponente: El Comité de Flora de CITES ha producido una serie de guías regionales con los nombres de personas expertas en CITES para contactar en los distintos países (véase la página web de CITES para más detalles). Posiblemente la guía le sirva a Ud. para establecer contacto con un experto relevante. Las características de orquídeas reproducidas artificialmente y de origen silvestre se explican en el Manual de Identificación CITES, Volumen 1, flora. La Secretaría CITES manda copias a todas las Autoridades CITES. Si su Autoridad no tiene una copia actualizada del Manual, póngase en contacto con la Secretaría CITES.]

Diapositiva 39

Diapositiva 39: *Paphiopedilum* y *Phragmipedium* ¿de origen silvestre o reproducidas artificialmente*?*

Esta diapositiva ilustra algunas de las características que pueden presentar las plantas reproducidas artificialmente y las de origen silvestre.

[Nota al ponente: Esta diapositiva muestra Paphiopedilum spp.]

Diapositiva 40: ***Phragmipedium*** **– posible comercio ilícito**

De nuevo, el posible comercio ilícito se centrará en las especies de más reciente descripción. Hemos resaltado el caso de *Phragmipedium kovachii*, cuyo descubrimiento suscitó gran interés en el mundo de las orquídeas, además de incitar al comercio ilícito. Esta especie sigue entre las más solicitadas, y es posible que aún se encuentren plantas de origen silvestre en el comercio ilícito durante algún tiempo en el futuro. La especie probablemente será objeto de contrabando, sin permiso o con una declaración errónea en el permiso.

Para ayudar en la comprobación de los nombres de las especies de *Phragmipedium* descritas más recientemente, sugerimos que los busque Ud. en las bases de datos que hemos señalado en la Sección de "Referencias y recursos".

[Nota al ponente: Esta diapositiva muestra tres imágenes de Phragmipedium kovachii.]

Diapositiva 41

Diapositiva 41: *Paphiopedilum* – posible comercio ilícito

Una vez más, el enfoque del posible comercio ilícito serán las especies más nuevas. Recientemente, las especies nuevas más interesantes se han encontrado en China y Vietnam. También se han descubierto nuevas especies en Filipinas, Indonesia y Malasia.

Hay que comprobar las importaciones procedentes de Asia. Además de las partidas comerciales, las posibles vías de contrabando de este material incluyen los enseres personales de viajeros internacionales, el equipaje de mano, el correo, y los servicios de mensajería.

Para ayudarle a verificar los nombres de especies de Paphiopedilum descritas recientemente, sugerimos que compruebe Ud. las bases de datos que hemos reseñado en la Sección de "Referencias y recursos". Ud. puede ver la fecha de publicación de un nuevo nombre de especie, y cuánto más reciente sea, más probable es que la planta sea de origen silvestre.

[Nota al ponente: Esta diapositiva muestra Paphiopedilum micranthum (izquierda) y Paphiopedilum armeniacum (derecha) en su hábitat natural de China – especies cuyas poblaciones silvestres tenían mucha demanda al principio de entrar en el comercio, en los años 1980.]

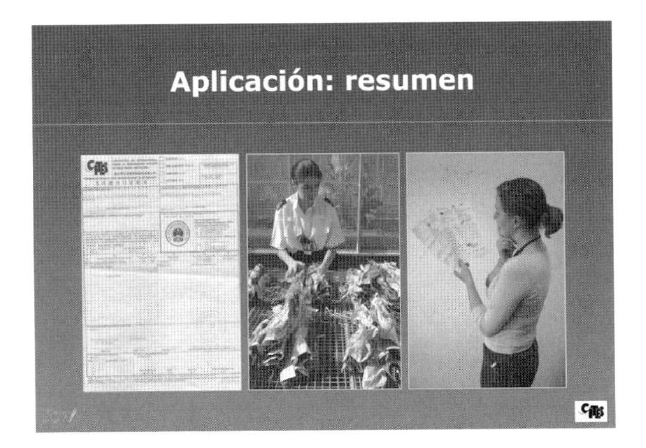

Diapositiva 42: Aplicación: resumen

Hemos cubierto los siguientes temas clave en la aplicación de CITES para las zapatillas de Venus:

- procedimientos de distintos países para hacer cumplir el Convenio;
- una lista de los aspectos a comprobar al efectuar las inspecciones;
- las características generales de las plantas reproducidas artificialmente y las de origen silvestre y;
- posible comercio ilícito.

Para ampliar la información sobre cuestiones de cumplimiento y formación, visite la página web de CITES: www.cites.org.

Diapositivas adicionales

Diapositiva 44

Diapositiva 44: Exenciones

A menudo, sólo las plantas reproducidas artificialmente entran en el comercio. Reconociendo este hecho, las Partes han decidido permitir algunas exenciones de los controles CITES en el caso de las orquídeas.

Para las plantas incluidas en el Apéndice I de CITES, se controlan tanto la planta entera como todas las partes y derivados, vivos o muertos. Sólo existe una exención, para cultivos de plántulas o de tejidos "obtenidos *in vitro*, en medios sólidos o líquidos, que se transportan en envases estériles". Este material, por supuesto, debe ser de origen legal para cumplir la exención.

En cuanto a las especies vegetales incluidas en el Apéndice II de CITES, se controla la planta "viva o muerta" y cualquier parte o derivado fácilmente identificable estipulado en los Apéndices. En el caso de *Cypripedium* las únicas partes y derivados exentos del control CITES son: a) las semillas y el polen (inclusive las polinias); b) los cultivos de plántulas o de tejidos obtenidos *in vitro*, en medios sólidos o líquidos, que se transportan en envases estériles; y c) las flores cortadas de ejemplares reproducidos artificialmente.

[Nota al ponente: Aunque numerosos híbridos de orquídeas están exentos de los controles CITES, las zapatillas de Venus no se encuentran entre ellos. Para ver los detalles sobre las exenciones aplicables a otros híbridos de orquídeas, visite la página web de CITES: www.cites.org.]

Diapositiva 45: Registro de viveros

La Resolución Conf. 9.19 (Rev. CoP13), *Directrices para el registro de viveros que exportan especímenes de especies incluidas en el Apéndice I reproducidos artificialmente*, establece los trámites del Convenio para registrar viveros. Fue adoptada en la 9ª Reunión de la Conferencia de las Partes en CITES (Fort Lauderdale, EE. UU., noviembre de 1994) y enmendada en la CdP13 (Bangkok, 2004). En CITES no se han determinado criterios para el registro de viveros que reproducen especímenes del Apéndice II, pero cualquier Autoridad nacional CITES tiene libertad para establecer un plan de registro de plantas del Apéndice II, por ejemplo, para simplificar la expedición de permisos. Un sistema así beneficiaría a los organismos y comerciantes locales, aunque el registro no tuviese valor fuera del país en cuestión.

La Autoridad Administrativa de cualquier Parte, previa consulta con la Autoridad Científica, puede remitir los datos de un determinado vivero a la Secretaría CITES, para su inscripción in el Registro de los establecimientos que reproducen especies vegetales del Apéndice I. El propietario del vivero, en primer lugar, debe presentar un perfil del establecimiento a la Autoridad Administrativa nacional. Dicho perfil incluirá, entre otras cosas, una descripción de las instalaciones, antecedentes y planes de reproducción, cantidad y tipo de plantas madre que mantiene, y pruebas de su adquisición legítima. La Autoridad Administrativa y la Autoridad Científica tienen que examinar estos datos y juzgar si el establecimiento reúne las cualidades para ser registrado. Durante estos trámites, sería razonable que las Autoridades nacionales efectuasen una inspección detallada del vivero.

Cuando las Autoridades nacionales estén satisfechas de la autenticidad del vivero y de la conveniencia de registrarlo, notificarán a la Secretaría CITES de esta opinión, dando detalles del vivero. La Autoridad Administrativa también debe describir el procedimiento de inspección que ha seguido para confirmar la identidad y el origen legal de los parentales de las plantas que van a entrar en el plan de registro, informando, asimismo, de cualquier otro material del Apéndice I presente en el vivero. Las Autoridades nacionales CITES además deberán velar por que no se agote el plantel reproductor de origen silvestre, y que se efectúe un estrecho control del establecimiento en general. Si se emplean semillas recolectadas del medio silvestre, la Autoridad Administrativa debe certificar que se ajusten a las condiciones expuestas en la definición CITES de "reproducida artificialmente" (véase la Diapositiva 46). La Autoridad Administrativa CITES también debe instaurar un sistema para agilizar la concesión de permisos y comunicar los pormenores del mismo a la Secretaría.

Si la Secretaría CITES está conforme con la información facilitada, entonces incluirá el vivero en su Registro de establecimientos. Si no, debe comunicar sus dudas a la Autoridad Administrativa, indicando los puntos que requieren aclaración. Cualquier Autoridad Administrativa CITES u otra fuente puede informar a la Secretaría del incumplimiento de los requisitos para registrarse. De sostenerse las dudas, el vivero puede ser suprimido del Registro, tras consultar con la Autoridad Administrativa.

Diapositiva 46: La definición CITES de "reproducida artificialmente"

La definición CITES de "reproducida artificialmente" está contenida en la Resolución Conf. 11.11 – *Reglamentación del comercio de plantas*. La definición dentro de CITES incluye algunos criterios únicos. Al aplicar estos criterios, se puede dar el caso de una planta que presenta todas las características físicas de la reproducción artificial, pero se considera de origen silvestre en términos de CITES. Los puntos clave son:

- *Las plantas deben cultivarse en un medio controlado.* Esto significa, por ejemplo, que las plantas se manipulen en *un medio no natural* con la finalidad de que las condiciones de cultivo sean las óptimas y para protegerlas contra predadores. Un vivero tradicional o un simple invernadero se considera "un medio controlado". Otro ejemplo de "un medio controlado" sería un umbráculo tropical con un buen plan de manejo o gestión. Lo que no constituye "un medio controlado" es la introducción eventual de un ejemplar de vegetación natural donde ya se dan especímenes silvestres de las plantas. Además, las plantas recolectadas en la naturaleza se consideran silvestres incluso si se han cultivado en condiciones controladas durante algún tiempo.

- El plantel parental cultivado debe haberse establecido *de forma que no sea perjudicial para la supervivencia de la especie en el medio silvestre* y gestionado *de tal manera que se garantice el mantenimiento a largo plazo de este plantel cultivado.*

- El plantel parental cultivado debe haberse establecido *con arreglo a las disposiciones de la CITES y de la legislación nacional correspondiente.* Esto quiere decir que el plantel debe haberse obtenido de forma legítima en términos de la CITES, y también según la legalidad vigente en el país de origen. Por ejemplo, es posible recolectar una planta de forma ilícita dentro de un país de origen, luego cultivarla en un vivero local y exportar su progenie declarada como reproducida artificialmente. Pero en términos de CITES no se puede considerar que dicha progenie haya sido reproducida artificialmente, debido a la recolección ilícita de las plantas madre.

Diapositiva 46

- Las semillas sólo pueden considerarse reproducidas artificialmente si se toman de plantas que también cumplen la definición CITES de "reproducida artificialmente". Se emplea el término *plantel parental cultivado* para permitir la incorporación de algunas plantas nuevas, extraídas del medio silvestre, para enriquecer el plantel reproductor. Se reconoce que pueden ser necesarios estos suplementos eventuales procedentes de la naturaleza, y esta práctica se permite, siempre que se haga de manera legítima y sostenible.

- Las plantas y las semillas se pueden considerar reproducidas artificialmente si se cultivan a partir de semillas recolectadas en el medio silvestre dentro de un determinado Estado del área de distribución, siempre y cuando dicha práctica cuente con la aprobación de las Autoridades Administrativa y Científica del país en cuestión.

Es complicado aplicar la definición CITES. Hay que comprobar el origen legal, la situación de la reproducción y la recolección no perjudicial de los especímenes. Para lograr una valoración correcta, hace falta la estrecha colaboración entre las Autoridades Administrativa y Científica CITES. En la práctica cotidiana, los criterios deben adaptarse a las circunstancias de cada una de las Partes en la Convención. Las Autoridades nacionales CITES deberían pensar en publicar una lista de los aspectos a tener en cuenta, como una manera de normalizar el proceso e informar a los comerciantes locales de plantas.

Diapositiva 47: Promover el comercio sostenible y el acceso a material para reproducción

La inclusión de grupos de plantas y animales en el Apéndice I de CITES en efecto prohíbe su intercambio con fines comerciales. La finalidad de la inclusión es proteger las plantas y los animales de un comercio perjudicial que podría llevarlos a la extinción. La inclusión en el Apéndice I no debe considerarse un éxito de conservación en sí. Más bien, un éxito de conservación es cuando ese taxón puede ser transferido al Apéndice II. Por lo tanto, es importante emprender acciones conservacionistas al incluirse un taxón en el Apéndice I de CITES. La demanda de taxa del Apéndice I no desaparece con su inclusión en dicho Apéndice, y teóricamente debe haber material reproducido artificialmente para satisfacer la demanda. Es posible disponer de este material cuando se han desarrollado las técnicas adecuadas de reproducción y cuando se cuenta con plantas madre de origen legal.

Sin embargo, en el caso de las zapatillas de Venus, nuevas especies se buscan, se encuentran, y se describen. A menudo es difícil encontrar plantas madre de origen legal para formar el plantel reproductor. Entonces el material ilegal se cuela en el mercado internacional y se incorpora paulatinamente en el plantel reproductor. Este proceso fomenta la recolección insostenible de las especies más escasas, privando al país de origen de los importantes ingresos que se podrían derivar de la introducción de este stock en el mercado internacional. Ha existido cierta inercia dentro de los países de origen, en el comercio internacional de orquídeas y en la comunidad CITES. Aguardamos una colaboración que logre crear mecanismos que permitan el acceso al material de reproducción y contribuyan a combatir el comercio ilícito. Los países de origen necesitan ayuda para establecer tales programas.

Es posible realizar actividades dentro de CITES para hacer que esto ocurra. Todo lo que se necesita es entusiasmo, iniciativa, confianza, y financiación. La confianza probablemente sea lo más difícil de conseguir. Estos programas de reproducción siempre serán vulnerables; podrán ser eclipsados por el comercio ilícito o incluso tachados de "bio-piratería". Si Ud. trabaja en CITES o en la industria de orquídeas, debe intentar alentar iniciativas de este tipo, pues son estos proyectos los que fomentan el comercio sostenible y facilitan el acceso de los países de origen a los fondos generados por sus propios recursos. También puede

Diapositiva 47

ser la única manera de afianzar las relaciones de asociación necesarias para la conservación de las especies y sus hábitats a largo plazo.

ÍNDICE

Slipper Orchid Names in Current Usage

Cypripedium**,** ***Paphiopedilum*** **and** ***Phragmipedium***

This list provides an update of the current accepted names in these three genera and is based on the World Checklist of Monocots (2005)* and revised following comments from experts. It is not an update of the CITES official checklist, which will be updated in the near future. It is an informal checklist- a work in progress!

***World Checklist of Monocots (2005) The Board of Trustees of the Royal Botanic Gardens, Kew. Published on the Internet; http://www.kew.org/monocotChecklist/ [accessed 3/2005].**

Noms utilisés actuellement pour les sabots de Vénus

Cypripedium*, *Paphiopedilum* et *Phragmipedium

Cette liste contient des informations mises à jour sur les noms acceptés actuellement pour ces trois genres, basées sur l'ouvrage World Checklist of Monocots (2005)* et révisées en fonction des premiers commentaires fournis par des experts. Merci de nous aider à finaliser le texte en nous envoyant vos commentaires. Cette liste n'est pas une version à jour de la liste de référence officielle de la CITES, qui sera mise à jour prochainement. Veuillez la considérer comme une liste de référence informelle – un travail en cours.

***World Checklist of Monocots (2005) Conseil d'administration, Royal Botanic Gardens, Kew. Publié sur Internet ; http://www.kew.org/monocotChecklist/ [consulté en mars 2005].**

Nombres de zapatillas de Venus actualmente en uso

Cypripedium*, *Paphiopedilum* y *Phragmipedium

La presente lista ofrece una puesta al día de los nombres actualmente aceptados en estos tres géneros, basada en la *World Checklist of Monocots* (*Lista de referencia mundial de monocotiledóneas*, 2005)* y revisada según comentarios iniciales de expertos. Por favor, ayúdenos a finalizar el texto enviándonos sus comentarios. No es una nueva versión de la lista de referencia oficial de CITES, que se actualizará en un futuro próximo. Por favor, trátela como una lista informal – ¡una obra en la que aún se está trabajando!

***World Checklist of Monocots (2005) The Board of Trustees of the Royal Botanic Gardens, Kew. Publicada en Internet; http://www.kew.org/monocotChecklist/ [vista en marzo de 2005].**

CYPRIPEDIUM – LIST OF ALL NAMES / CYPRIPEDIUM – LISTE COMPLÈTE DES NOMS / CYPRIPEDIUM – LISTA DE TODOS LOS NOMBRES

Accepted names are given in **bold** / Les noms acceptés sont en **caractères gras** / Los nombres aceptados se dan en **negrita**

ALL NAMES TOUS LES NOMS TODOS LOS NOMBRES	ACCEPTED NAME NOM ACCEPTÉS NOMBRES ACEPTADOS
Cypripedium acaule	
Cypripedium acaule f. *biflorum*	**Cypripedium acaule**
Cypripedium acaule var. *album*	**Cypripedium acaule**
Cypripedium album	**Cypripedium reginae**
Cypripedium alternifolium	**Cypripedium calceolus**
Cypripedium amesianum	**Cypripedium yunnanense**
Cypripedium appletonianum	**Paphiopedilum appletonianum**
Cypripedium argus	**Paphiopedilum argus**
Cypripedium arietinum	
Cypripedium arietinum f. *biflorum*	**Cypripedium arietinum**
Cypripedium assamicum	**Paphiopedilum fairrieanum**
Cypripedium assurgens	**Cypripedium parviflorum var. pubescens**
Cypripedium atsmori	**Cypripedium macranthos**
Cypripedium aureum	**Cypripedium parviflorum var. pubescens**
Cypripedium barbatum	**Paphiopedilum barbatum**
Cypripedium barbatum var. *biflorum*	**Paphiopedilum barbatum**
Cypripedium barbatum var. *crossi*	**Paphiopedilum callosum**
Cypripedium barbatum var. *superbum*	**Paphiopedilum superbiens**
Cypripedium barbatum var. *veitchii*	**Paphiopedilum superbiens**
Cypripedium barbatum var. *warneri*	**Paphiopedilum callosum var. sublaeve**
Cypripedium barbatum var. warnerianum	**Paphiopedilum callosum var. warnerianum**
Cypripedium bardolphianum	
Cypripedium bardolphianum var. *zhongdianense*	**Cypripedium forrestii**
Cypripedium bellatulum	**Paphiopedilum bellatulum**
Cypripedium bifidum	**Cypripedium parviflorum**
Cypripedium biflorum	**Paphiopedilum barbatum**
Cypripedium binotii	**Phragmipedium vittatum**
Cypripedium boissierianum	**Phragmipedium boissierianum**
Cypripedium boreale	**Cypripedium calceolus**
Cypripedium boxallii	**Paphiopedilum villosum var. boxallii**
Cypripedium boxallii var. *atratum*	**Paphiopedilum villosum var. boxallii**
Cypripedium bulbosum	**Calypso bulbosa**
Cypripedium bulbosum var. *parviflorum*	**Cypripedium parviflorum**
Cypripedium bullenianum	**Paphiopedilum bullenianum**
Cypripedium bullenianum var. *appletonianum*	**Paphiopedilum appletonianum**
Cypripedium calceolus	
Cypripedium calceolus var. *delta*	**Cypripedium guttatum**
Cypripedium calceolus var. *gamma*	**Cypripedium reginae**
Cypripedium calceolus var. *parviflorum*	**Cypripedium parviflorum**
Cypripedium calceolus var. *planipetalum*	**Cypripedium parviflorum var. pubescens**
Cypripedium calceolus var. *pubescens*	**Cypripedium parviflorum var. pubescens**
Cypripedium calcicolum	

ALL NAMES	ACCEPTED NAMES
Cypripedium californicum	
Cypripedium callosum	**Paphiopedilum callosum**
Cypripedium callosum var. *sublaeve*	**Paphiopedilum callosum var. warnerianum**
Cypripedium canadense	**Cypripedium reginae**
Cypripedium candidum	
Cypripedium cannartianum	**Paphiopedilum philippinense**
Cypripedium cardiophyllum	**Cypripedium debile**
Cypripedium caricinum	**Phragmipedium caricinum**
Cypripedium caricinum Bateman	**Phragmipedium caricinum**
Cypripedium cathayenum	**Cypripedium japonicum**
Cypripedium caudatum	**Phragmipedium caudatum**
Cypripedium caudatum var. *lindenii*	**Phragmipedium lindenii subsp. lindenii**
Cypripedium caudatum var. *wallisii*	**Phragmipedium lindenii subsp. wallisii**
Cypripedium caudatum var. warscewiczii	**Phragmipedium exstaminodium subsp. warscewiczii**
Cypripedium chamberlainianum	**Paphiopedilum victoria-regina**
Cypripedium chantinii	**Paphiopedilum insigne**
Cypripedium charlesworthii	**Paphiopedilum charlesworthii**
Cypripedium cheniae	**Cypripedium fasciolatum**
Cypripedium chica	**Selenipedium chica**
Cypripedium chinense	**Cypripedium henryi**
Cypripedium ciliolare	**Paphiopedilum ciliolare**
Cypripedium ciliolare var. *miteauanum*	**Paphiopedilum ciliolare**
Cypripedium compactum	**Cypripedium tibeticum**
Cypripedium concolor	**Paphiopedilum concolor**
Cypripedium concolor var. *chlorophyllum*	**Paphiopedilum concolor**
Cypripedium concolor var. *godefroyae*	**Paphiopedilum godefroyae**
Cypripedium concolor var. *reynieri*	**Paphiopedilum concolor**
Cypripedium concolor var. *sulphurinum*	**Paphiopedilum concolor**
Cypripedium concolor var. *tonkinense*	**Paphiopedilum concolor**
Cypripedium cordigerum	
Cypripedium corrugatum	**Cypripedium tibeticum**
Cypripedium corrugatum var. *obesum*	**Cypripedium tibeticum**
Cypripedium cothurnum	**Catasetum macrocarpum**
Cypripedium crawshayae	**Paphiopedilum charlesworthii**
Cypripedium crossii	**Paphiopedilum barbatum**
Cypripedium cruciatum	**Cypripedium calceolus**
Cypripedium cruciforme	**Paphiopedilum lowii**
Cypripedium curtisii	**Paphiopedilum superbiens var. curtisii**
Cypripedium curtisii var. *sanderae*	**Paphiopedilum superbiens var. curtisii**
Cypripedium czerwiakowianum	**Phragmipedium boissierianum var. czerwiakowianum**
Cypripedium daliense	**Cypripedium margaritaceum**
Cypripedium dariense	**Phragmipedium longifolium**
Cypripedium daultonii	**Cypripedium kentuckiense**
Cypripedium dauthieri	**Paphiopedilum hennisianum**
Cypripedium dayanum	**Paphiopedilum dayanum**
Cypripedium debile	
Cypripedium delenatii	**Paphiopedilum delenatii**
Cypripedium dickinsonianum	
Cypripedium dilectum	**Paphiopedilum villosum var. boxallii**
Cypripedium dominianum	**Phragmipedium Dominianum (hybrid)**
Cypripedium druryi	**Paphiopedilum druryi**
Cypripedium ebracteatum	**Cypripedium fargesii**
Cypripedium elatum	**Phragmipedium humboldtii**

ALL NAMES	ACCEPTED NAMES
Cypripedium elegans	
Cypripedium elliottianum	**Paphiopedilum rothschildianum**
Cypripedium epidendricum	**Eulophia alta**
Cypripedium ernestianum	**Paphiopedilum dayanum**
Cypripedium exul	**Paphiopedilum exul**
Cypripedium fairrieanum	**Paphiopedilum fairrieanum**
Cypripedium fargesii	
Cypripedium farreri	
Cypripedium fasciculatum	
Cypripedium fasciculatum var. *pusillum*	**Cypripedium fasciculatum**
Cypripedium fasciolatum	
Cypripedium ferrugineum	**Cypripedium calceolus**
Cypripedium flavescens	**Cypripedium parviflorum var. pubescens**
Cypripedium flavum	
Cypripedium formosanum	
Cypripedium forrestii	
Cypripedium franchetii	
Cypripedium froschii	**Cypripedium x froschii (hybrid)**
Cypripedium furcatum	**Cypripedium parviflorum var. pubescens**
Cypripedium gardineri	**Paphiopedilum glanduliferum**
Cypripedium glanduliferum	**Paphiopedilum glanduliferum**
Cypripedium glaucophyllum	**Cypripedium glaucophyllum**
Cypripedium godefroyae	**Paphiopedilum godefroyae**
Cypripedium godefroyae var. leucochilum	**Paphiopedilum godefroyae**
Cypripedium grandiflorum	**Phragmipedium boissierianum**
Cypripedium gratrixianum	**Paphiopedilum gratrixianum**
Cypripedium guttatum	
Cypripedium guttatum f. *albiflorum*	**Cypripedium guttatum**
Cypripedium guttatum subsp. *yatabeanum*	**Cypripedium yatabeanum**
Cypripedium guttatum var. *segawai*	**Cypripedium segawai**
Cypripedium guttatum var. *wardii*	**Cypripedium wardii**
Cypripedium guttatum var. *yatabeanum*	**Cypripedium yatabeanum**
Cypripedium hartwegii	**Phragmipedium longifolium**
Cypripedium haynaldianum	**Paphiopedilum haynaldianum**
Cypripedium henryi	
Cypripedium himalaicum	
Cypripedium hincksianum	**Phragmipedium longifolium**
Cypripedium hirsutissimum	**Paphiopedilum hirsutissimum**
Cypripedium hirsutum	**Cypripedium acaule**
Cypripedium hookerae	**Paphiopedilum hookerae**
Cypripedium hookerae var. *volonteanum*	**Paphiopedilum hookerae var. volonteanum**
Cypripedium humboldtii	**Phragmipedium exstaminodium subsp. warscewiczii**
Cypripedium humile	**Cypripedium acaule**
Cypripedium hyeanum	**Paphiopedilum lawrenceanum**
Cypripedium insigne	**Paphiopedilum insigne**
Cypripedium insigne var. *exul*	**Paphiopedilum exul**
Cypripedium insigne var. *sanderae*	**Paphiopedilum insigne**
Cypripedium insigne var. *sanderianum*	**Paphiopedilum insigne**
Cypripedium irapeanum	
Cypripedium isabelianum	**Selenipedium isabelianum**
Cypripedium japonicum	
Cypripedium japonicum var. *formosanum*	**Cypripedium formosanum**
Cypripedium japonicum var. *glabrum*	**Cypripedium japonicum**
Cypripedium javanicum	**Paphiopedilum javanicum**
Cypripedium javanicum var. *virens*	**Paphiopedilum javanicum var. virens**
Cypripedium kentuckiense	
Cypripedium kentuckiense f. *pricei*	**Cypripedium kentuckiense**

ALL NAMES	ACCEPTED NAMES
Cypripedium klotzschianum	**Phragmipedium klotzschianum**
Cypripedium knightae	**Cypripedium fasciculatum**
Cypripedium laevigatum	**Paphiopedilum philippinense**
Cypripedium langrhoa	**Cypripedium tibeticum**
Cypripedium lanuginosum	**Cypripedium franchetii**
Cypripedium lawrenceanum	**Paphiopedilum lawrenceanum**
Cypripedium lawrenceanum var. *hyeanum*	**Paphiopedilum lawrenceanum**
Cypripedium lentiginosum	
Cypripedium lexarzae	**Cypripedium irapeanum**
Cypripedium lichiangense	
Cypripedium lindenii	**Phragmipedium lindenii var. lindenii**
Cypripedium lindleyanum	**Phragmipedium lindleyanum**
Cypripedium linearisubulatum	**Cleisostoma subulatum**
Cypripedium longifolium	**Phragmipedium longifolium**
Cypripedium longifolium var. *gracile*	**Phragmipedium longifolium**
Cypripedium lowii	**Paphiopedilum lowii**
Cypripedium ludlowii	
Cypripedium luteum franch.	**Cypripedium flavum**
Cypripedium luteum Raf.	**Cypripedium parviflorum**
Cypripedium luteum var. *angustifolium*	**Cypripedium parviflorum var. pubescens**
Cypripedium luteum var. *biflorum*	**Cypripedium parviflorum var. pubescens**
Cypripedium luteum var. *concolor*	**Cypripedium parviflorum var. pubescens**
Cypripedium luteum var. *glabrum*	**Cypripedium parviflorum var. pubescens**
Cypripedium luteum var. *grandiflorum*	**Cypripedium parviflorum var. pubescens**
Cypripedium luteum var. *maculatum*	**Cypripedium parviflorum var. pubescens**
Cypripedium luteum var. *parviflorum*	**Cypripedium parviflorum**
Cypripedium luteum var. *pubescens*	**Cypripedium parviflorum var. pubescens**
Cypripedium luzmarianum	**Cypripedium molle**
Cypripedium macranthos	
Cypripedium macranthos nothof. *alboroseum*	**Cypripedium macranthos**
Cypripedium macranthos nothof. *albostriatum*	**Cypripedium macranthos**
Cypripedium macranthos nothof. *flavoroseum*	**Cypripedium macranthos**
Cypripedium macranthos var. *album*	**Cypripedium macranthos**
Cypripedium macranthos var. *atropurpureum*	**Cypripedium macranthos**
Cypripedium macranthos var. *flavum*	**Cypripedium macranthos**
Cypripedium macranthos var. *himalaicum*	**Cypripedium himalaicum**
Cypripedium macranthos var. *tibeticum*	**Cypripedium tibeticum**
Cypripedium macranthos var. *ventricosum*	**Cypripedium x ventricosum (hybrid)**
Cypripedium macranthos var. *villosum*	**Cypripedium franchetii**
Cypripedium makasin	**Cypripedium parviflorum var. makasin**
Cypripedium margaritaceum	
Cypripedium margaritaceum var. *fargesii*	**Cypripedium fargesii**
Cypripedium marianus	**Cypripedium calceolus**
Cypripedium mastersianum	**Paphiopedilum mastersianum**
Cypripedium maulei	**Paphiopedilum insigne**
Cypripedium micranthum	
Cypripedium microsaccos	**Cypripedium calceolus**
Cypripedium miteauanum	**Paphiopedilum ciliolare**
Cypripedium moensianum	**Paphiopedilum argus**
Cypripedium moensii	**Paphiopedilum argus**
Cypripedium molle	
Cypripedium montanum	

ALL NAMES	ACCEPTED NAMES
Cypripedium montanum f. *praetertinctum*	**Cypripedium montanum**
Cypripedium montanum f. *welchii*	**Cypripedium montanum**
Cypripedium neoguineense	**Paphiopedilum rothschildianum**
Cypripedium nigritum	**Paphiopedilum barbatum**
Cypripedium niveum	**Paphiopedilum niveum**
Cypripedium niveum var. *album*	**Paphiopedilum niveum**
Cypripedium nutans	**Cypripedium bardolphianum**
Cypripedium occidentale	**Cypripedium montanum**
Cypripedium orientale	**Cypripedium guttatum**
Cypripedium palangshanense	
Cypripedium palmifolium	**Selenipedium palmifolium**
Cypripedium papuanum	**Paphiopedilum papuanum**
Cypripedium pardinum	**Paphiopedilum venustum**
Cypripedium parishii	**Paphiopedilum parishii**
Cypripedium parviflorum	
Cypripedium parviflorum f. *albolabium*	**Cypripedium parviflorum**
Cypripedium parviflorum var. makasin	
Cypripedium parviflorum var. *planipetalum*	**Cypripedium parviflorum var. pubescens**
Cypripedium parviflorum var. pubescens	
Cypripedium passerinum	
Cypripedium passerinum var. *minganense*	**Cypripedium passerinum**
Cypripedium paulistanum	**Phragmipedium vittatum**
Cypripedium pearcei	**Phragmipedium pearcei**
Cypripedium petri	**Paphiopedilum dayanum**
Cypripedium philippinense	**Paphiopedilum philippinense**
Cypripedium philippinense var. *roebelenii*	**Paphiopedilum philippinense var. roebelenii**
Cypripedium pitcherianum	**Paphiopedilum argus**
Cypripedium planipetalum	**Cypripedium parviflorum var. pubescens**
Cypripedium platytaenium	**Paphiopedilum stonei var. platytaenium**
Cypripedium plectrochilum	
Cypripedium praestans	**Paphiopedilum glanduliferum**
Cypripedium praestans var. *kimballianum*	**Paphiopedilum glanduliferum**
Cypripedium pubescens	**Cypripedium parviflorum var. pubescens**
Cypripedium pubescens var. makasin	**Cypripedium parviflorum var. makasin**
Cypripedium pulchrum	**Cypripedium franchetii**
Cypripedium purpuratum	**Paphiopedilum purpuratum**
Cypripedium purpuratum	**Paphiopedilum barbatum**
Cypripedium pusillum	**Cypripedium fasciculatum**
Cypripedium reginae	
Cypripedium reginae var. *album*	**Cypripedium reginae**
Cypripedium reichenbachianum	**Phragmipedium longifolium**
Cypripedium reichenbachii	**Phragmipedium longifolium**
Cypripedium reticulatum	**Phragmipedium reticulatum**
Cypripedium robinsonii	**Paphiopedilum bullenianum**
Cypripedium roebelenii	**Paphiopedilum philippinense var. roebelenii**
Cypripedium roebelenii var. *cannartianum*	**Paphiopedilum philippinense**
Cypripedium roezlii	**Phragmipedium longifolium**
Cypripedium rothschildianum	**Paphiopedilum rothschildianum**
Cypripedium rubronerve	**Cypripedium x ventricosum (hybrid)**
Cypripedium sanderianum	**Paphiopedilum sanderianum**
Cypripedium sargentianum	**Phragmipedium lindleyanum**
Cypripedium schlimii	**Phragmipedium schlimii**
Cypripedium schlimii var. *albiflorum*	**Phragmipedium schlimii**
Cypripedium schmidtianum	**Paphiopedilum callosum**

ALL NAMES	ACCEPTED NAMES
Cypripedium schomburgkianum	**Phragmipedium klotzschianum**
Cypripedium segawai	
Cypripedium shanxiense	
Cypripedium sinicum	**Paphiopedilum purpuratum**
Cypripedium smithii	**Cypripedium calcicolum**
Cypripedium socco	**Catasetum socco**
Cypripedium speciosum	**Cypripedium macranthos**
Cypripedium spectabile	**Cypripedium reginae**
Cypripedium spectabile var. *dayanum*	**Paphiopedilum dayanum**
Cypripedium spicerianum	**Paphiopedilum spicerianum**
Cypripedium splendidum	**Cypripedium irapeanum**
Cypripedium stonei	**Paphiopedilum stonei**
Cypripedium subtropicum	
Cypripedium superbiens	**Paphiopedilum superbiens**
Cypripedium superbiens var. *dayanum*	**Paphiopedilum dayanum**
Cypripedium taibaiense	**Cypripedium calcicolum**
Cypripedium taiwanianum	**Cypripedium macranthos**
Cypripedium thunbergii	**Cypripedium macranthos**
Cypripedium tibeticum	
Cypripedium tonkinense	**Paphiopedilum concolor**
Cypripedium tonsum	**Paphiopedilum tonsum**
Cypripedium turgidum	**Cypripedium irapeanum**
Cypripedium undatum	**Cypripedium parviflorum var. pubescens**
Cypripedium variegatum	**Cypripedium guttatum**
Cypripedium veganum	**Cypripedium parviflorum var. pubescens**
Cypripedium veitchianum	**Paphiopedilum superbiens**
Cypripedium venustum	**Paphiopedilum venustum**
Cypripedium venustum var. *measuresianum*	**Paphiopedilum venustum**
Cypripedium vernayi	**Paphiopedilum wardii**
Cypripedium victoria-mariae	**Paphiopedilum victoria-mariae**
Cypripedium victoria-regina	**Paphiopedilum victoria-regina**
Cypripedium villosum	**Paphiopedilum villosum**
Cypripedium villosum var. *boxallii*	**Paphiopedilum villosum var. boxallii**
Cypripedium virens	**Paphiopedilum javanicum var. virens**
Cypripedium vittatum	**Phragmipedium vittatum**
Cypripedium volonteanum	**Paphiopedilum hookerae var. volonteanum**
Cypripedium wallisii	**Phragmipedium lindenii var. wallisii**
Cypripedium waltersianum	**Paphiopedilum appletonianum**
Cypripedium wardianum	**Paphiopedilum wardii**
Cypripedium wardii	
Cypripedium wardii	**Paphiopedilum wardii**
Cypripedium warnerianum	**Paphiopedilum barbatum**
Cypripedium warsczewiczianum	**Phragmipedium caudatum**
Cypripedium wilsonii	**Cypripedium fasciculatum**
Cypripedium wolterianun	**Paphiopedilum appletonianum**
Cypripedium wumengense	
Cypripedium yatabeanum	
Cypripedium yunnanense	
Cypripedium zhongdianense	**Cypripedium farreri**

PAPHIOPEDILUM – LIST OF ALL NAMES / PAPHIOPEDILUM – LISTE COMPLÈTE DES NOMS / PAPHIOPEDILUM – LISTA DE TODOS LOS NOMBRES

ALL NAMES TOUS LES NOMS TODOS LOS NOMBRES	ACCEPTED NAME NOM ACCEPTÉS NOMBRES ACEPTADOS
Paphiopedilum acmodontum	
Paphiopedilum adductum	
Paphiopedilum adductum var. *anitum*	**Paphiopedilum adductum**
Paphiopedilum aestivum	**Paphiopedilum purpuratum**
Paphiopedilum amabile	**Paphiopedilum bullenianum**
Paphiopedilum ambonensis	**Paphiopedilum bullenianum var. celebesense**
Paphiopedilum ang-thong	**Paphiopedilum godefroyae**
Paphiopedilum angustatum	**Paphiopedilum malipoense var. hiepii**
Paphiopedilum angustifolium	**Paphiopedilum appletonianum**
Paphiopedilum anitum	**Paphiopedilum adductum**
Paphiopedilum appletonianum	
Paphiopedilum appletonianum f. *album*	**Paphiopedilum appletonianum**
Paphiopedilum appletonianum f. *immaculatum*	**Paphiopedilum appletonianum**
Paphiopedilum appletonianum var. *hainanense*	**Paphiopedilum appletonianum**
Paphiopedilum appletonianum var. *immaculatum*	**Paphiopedilum appletonianum**
Paphiopedilum appletonianum var. *poyntziamum*	**Paphiopedilum appletonianum**
Paphiopedilum argus	
Paphiopedilum argus var. *sriwanae*	**Paphiopedilum argus**
Paphiopedilum armeniacum	
Paphiopedilum armeniacum f. *markii*	**Paphiopedilum armeniacum**
Paphiopedilum armeniacum var. *mark-fun*	**Paphiopedilum armeniacum**
Paphiopedilum armeniacum var. *markii*	**Paphiopedilum armeniacum**
Paphiopedilum armeniacum var. *parviflorum*	**Paphiopedilum rmeniacum**
Paphiopedilum armeniacum var. *undulatum*	**Paphiopedilum armeniacum**
Paphiopedilum bacanum	**Paphiopedilum schoseri**
Paphiopedilum barbatum	
Paphiopedilum barbatum subsp. *lawrenceanum*	**Paphiopedilum barbatum**
Paphiopedilum barbatum var. *argus*	**Paphiopedilum argus**
Paphiopedilum barbatum var. *hennisianum*	**Paphiopedilum barbatum**
Paphiopedilum barbatum var. *nigritum*	**Paphiopedilum barbatum**
Paphiopedilum barbigerum	
Paphiopedilum barbigerum f. *aureum*	**Paphiopedilum barbigerum**
Paphiopedilum barbigerum var. *aureum*	**Paphiopedilum barbigerum**
Paphiopedilum bellatulum	
Paphiopedilum bellatulum f. *album*	**Paphiopedilum bellatulum**
Paphiopedilum bellatulum var. *album*	**Paphiopedilum bellatulum**
Paphiopedilum besseae	**Phragmipedium besseae**
Paphiopedilum birkii	**Paphiopedilum callosum var. sublaeve**
Paphiopedilum bodegomii	**Paphiopedilum wilhelminae**
Paphiopedilum boissierianum	**Phragmipedium boissierianum**
Paphiopedilum bougainvilleanum	
Paphiopedilum bougainvilleanum var. *saskianum*	**Paphiopedilum bougainvilleanum**
Paphiopedilum boxallii	**Paphiopedilum villosum var. boxallii**
Paphiopedilum braemii	**Paphiopedilum tonsum var. braemii**
Pahiopedilum brevilabium	**Paphiopedilum wardii**
Paphiopedilum bullenianum	
Paphiopedilum bullenianum var. bullenianum	
Paphiopedilum bullenianum var. celebesense	

ALL NAMES	ACCEPTED NAMES
Paphiopedilum bullenianum var. *johorense*	**Paphiopedilum bullenianum**
Paphiopedilium burmanicum	**Paphiopedilum wardii**
Paphiopedilum callosum	
Paphiopedilum callosum subsp. *sublaeve*	**Paphiopedilum callosum**
Paphiopedilum callosum var. *angustipetalum*	**Paphiopedilum callosum**
Paphiopedilum callosum var. callosum	
Paphiopedilum callosum var. potentianum	
Paphiopedilum callosum var. *schmidtianum*	**Paphiopedilum callosum**
Paphiopedilum callosum var. *sublaeve*	**Paphiopedilum callosum**
Paphiopedilum callosum var. *warnerianum*	**Paphiopedilum callosum**
Paphiopedilum caobangense	**Paphiopedilum tranlienianum**
Paphiopedilum caricinum	**Phragmipedium caricinum**
Paphiopedilum caudatum	**Phragmipedium caudatum**
Paphiopedilum caudatum var. *lindenii*	**Phragmipedium lindenii**
Paphiopedilum caudatum var. *wallisii*	**Phragmipedium wallisii**
Paphiopedilum celebesense	**Paphiopedilum bullenianum var. celebesense**
Paphiopedilum ceramensis	**Paphiopedilum bullenianum var. celebesense**
Paphiopedilum cerveranum	**Paphiopedilum appletonianum**
Paphiopedilum cerveranum f. *viride*	**Paphiopedilum appletonianum**
Paphiopedilum chamberlainianum	**Paphiopedilum victoria-regina**
Paphiopedilum chamberlainianum f. *primulinum*	**Paphiopedilum primulinum**
Paphiopedilum chamberlainianum f. *victoria-mariae*	**Paphiopedilum victoria-mariae**
Paphiopedilum chamberlainianum subsp. *liemianum*	**Paphiopedilum liemianum**
Paphiopedilum chamberlainianum var. *flavescens*	**Paphiopedilum primulinum var. purpurascens**
Paphiopedilum chamberlainianum var. *flavum*	**Paphiopedilum primulinum**
Paphiopedilum chamberlainianum var. *liemianum*	**Paphiopedilum liemianum**
Paphiopedilum chamberlainianum var. *primulinum*	**Paphiopedilum primulinum**
Paphiopedilum chaoi	**Paphiopedilum henryanum**
Paphiopedilum charlesworthii	
Paphiopedilum charlesworthii f. *crawshayae*	**Paphiopedilum charlesworthii**
Paphiopedilum charlesworthii f. *sandowiae*	**Paphiopedilum charlesworthii**
Paphiopedilum chiwuanum	**Paphiopedilum hirsutissimum var. chiwuanum**
Paphiopedilum ciliolare	
Paphiopedilum ciliolare var. *miteauanum*	**Paphiopedilum ciliolare**
Paphiopedilum coccineum	
Paphiopedilum concolor	
Paphiopedilum concolor f. *sulphurinum*	**Paphiopedilum concolor**
Paphiopedilum concolor subsp. *chlorophyllum*	**Paphiopedilum concolor**
Paphiopedilum concolor subsp. *reynieri*	**Paphiopedilum concolor**
Paphiopedilum concolor var. *dahuaense*	**Paphiopedilum concolor**
Paphiopedilum concolor var. *immaculatum*	**Paphiopedilum concolor**
Paphiopedilum concolor var. *niveum*	**Paphiopedilum niveum**
Paphiopedilum cothurnum	**Catasetum macrocarpum**
Paphiopedilum crossii	**Paphiopedilum barbatum**
Paphiopedilum crossii var. *potentianum*	**Paphiopedilum callosum var. potentianum**
Paphiopedilum crossii var. *sublaeve*	**Paphiopedilum callosum var. warnerianum**
Paphiopedilum curtisii	**Paphiopedilum superbiens**
Paphiopedilum czerwiakowianum	**Phragmipedium boissierianum var. czerwiakowianum**
Paphiopedilum dariense	**Phragmipedium longifolium**
Paphiopedilum dayanum	
Paphiopedilum dayanum var. *petri*	**Paphiopedilum dayanum**
Paphiopedilum delenatii	
Paphiopedilum delenatii f. *albinum*	**Paphiopedilum delenatii**
Paphiopedilum delicatum	**Paphiopedilum helenae**

ALL NAMES	ACCEPTED NAMES
Paphiopedilum dennisii	**Paphiopedilum wentworthianum**
Paphiopedilum densissimum	**Paphiopedilum villosum**
Paphiopedilum devogelii	**Paphiopedilum supardii**
Paphiopedilum dianthum	
Paphiopedilum dilectum	**Paphiopedilum villosum var. boxallii**
Paphiopedilum dollii	**Paphiopedilum henryanum**
Paphiopedilum druryi	
Paphiopedilum ecuadorense	**Phragmipedium pearcei**
Paphiopedilum elliottianum	**Paphiopedilum rothschildianum**
Paphiopedilum elliottianum sensu Fowlie	**Paphiopedilum adductum**
Paphiopedilum emersonii	
Paphiopedilum emersonii f. *luteum*	**Paphiopedilum emersonii**
Paphiopedilum epidendricum	**Eulophia alta**
Paphiopedilum esquirolei	**Paphiopedilum hirsutissimum var. esquirolei**
Paphiopedilum exstaminodium	**Phragmipedium exstaminodium**
Paphiopedilum exul	
Paphiopedilum fairrieanum	
Paphiopedilum fairrieanum f. *bohlmannianum*	**Paphiopedilum fairrieanum**
Paphiopedilum fairrieanum var. *bohlmannianum*	**Paphiopedilum fairrieanum**
Paphiopedilum fairrieanum var. *giganteum*	**Paphiopedilum fairrieanum**
Paphiopedilum fairrieanum var. *nigrescens*	**Paphiopedilum fairrieanum**
Paphiopedilum fowliei	
Paphiopedilum fowliei f. *christianae*	**Paphiopedilum fowliei**
Paphiopedilum fowliei f. *sangianum*	**Paphiopedilum fowliei**
Paphiopedilum fowliei var. *sangianum*	**Paphiopedilum fowliei**
Paphiopedilum gardineri Guillemard	
Paphiopedilum gardineri sensu Kennedy non Guillemard	**Paphiopedilum glanduliferum var. wilhelminae**
Paphiopedilum gigantifolium	
Paphiopedilum glanduliferum	
Paphiopedilum glanduliferum var. *gardineri*	**Paphiopedilum glanduliferum**
Paphiopedilum glanduliferum var. *kimballianum*	**Paphiopedilum glanduliferum**
Paphiopedilum glanduliferum var. *praestans*	**Paphiopedilum glanduliferum**
Paphiopedilum glanduliferum var. *wilhelminae*	**Paphiopedilum wilhelminae**
Paphiopedilum glaucophyllum	
Paphiopedilum glaucophyllum f. *flavoviride*	**Paphiopedilum glaucophyllum**
Paphiopedilum glaucophyllum var. glaucophyllum	
Paphiopedilum glaucophyllum var. moquetteanum	
Paphiopedilum globulosum	**Paphiopedilum micranthum**
Paphiopedilum godefroyae	
Paphiopedilum godefroyae var. leucochilum	
Paphiopedilum godefroyae var. *ang-thong*	**Paphiopedilum godefroyae**
Paphiopedilum gratrixianum	
Paphiopedilum hainanense	**Paphiopedilum appletonianum**
Paphiopedilum hangianum	
Paphiopedilum hartwegii	**Phragmipedium longifolium**
Paphiopedilum haynaldianum	
Paphiopedilum haynaldianum f. *album*	**Paphiopedilum haynaldianum**
Paphiopedilum helenae	
Paphiopedilum helenae f. *aureum*	**Paphiopedilum helenae**
Paphiopedilum hennisianum	
Paphiopedilum hennisianum f. *christiansenii*	**Paphiopedilum hennisianum**
Paphiopedilum hennisianum var. *christiansenii*	**Paphiopedilum hennisianum**
Paphiopedilum hennisianum var. *fowliei*	**Paphiopedilum fowliei**
Paphiopedilum henryanum	
Paphiopedilum henryanum f. *christae*	**Paphiopedilum henryanum**
Paphiopedilum henryanum var. *christae*	**Paphiopedilum henryanum**
Paphiopedilum herrmannii	
Paphiopedilum hiepii	**Paphiopedilum jackii var. hiepii**
Paphiopedilum hilmarii	**Paphiopedilum vietnamense**

ALL NAMES	ACCEPTED NAMES
Paphiopedilum hincksianum	**Phragmipedium longifolium**
Paphiopedilum hirsutissimum	
Paphiopedilum hirsutissimum f. *viride*	**Paphiopedilum hirsutissimum var. esquirolei**
Paphiopedilum hirsutissimum var. chiwuanum	
Paphiopedilum hirsutissimum var. esquirolei	
Paphiopedilum hirtzii	**Phragmipedium hirtzii**
Paphiopedilum hookerae	
Paphiopedilum hookerae f. *sandowiae*	**Paphiopedilum hookerae var. volonteanum**
Paphiopedilum hookerae subsp. *appletonianum*	**Paphiopedilum appletonianum**
Paphiopedilum hookerae var. *bullenianum*	**Paphiopedilum bullenianum**
Paphiopedilum hookerae var. volonteanum	
Paphiopedilum huonglanae	**Paphiopedilum emersonii**
Paphiopedilum insigne	
Paphiopedilum insigne f. *sanderae*	**Paphiopedilum insigne**
Paphiopedilum insigne f. *sanderianum*	**Paphiopedilum insigne**
Paphiopedilum insigne var. *barbigerum*	**Paphiopedilum barbigerum**
Paphiopedilum insigne var. *exul*	**Paphiopedilum exul**
Paphiopedilum intaniae	
Paphiopedilum jackii	
Paphiopedilum jackii var. hiepii	
Paphiopedilum javanicum	
Paphiopedilum javanicum f. *nymphenburgianum*	**Paphiopedilum javanicum**
Paphiopedilum javanicum var. *nymphenburgianum*	**Paphiopedilum javanicum**
Paphiopedilum javanicum var. virens	
Paphiopedilum johorense	**Paphiopedilum bullenianum**
Paphiopedilum kaieteurum	**Phragmipedium lindleyanum**
Paphiopedilum kalinae	**Paphiopedilum victoria-regina**
Paphiopedilum klotzschianum	**Phragmipedium klotzschianum**
Paphiopedilum kolopakingii	
Paphiopedilum laevigatum	**Paphiopedilum philippinense**
Paphiopedilum lawrenceanum	
Paphiopedilum lawrenceanum f. *hyeanum*	**Paphiopedilum lawrenceanum**
Paphiopedilum lawrenceanum var. *hyeanum*	**Paphiopedilum lawrenceanum**
Paphiopedilum leucochilum	**Paphiopedilum godefroyae var. leucochilum**
Paphiopedilum liemianum	
Paphiopedilum liemianum f. *purpurascens*	**Paphiopedilum primulinum var. purpurascens**
Paphiopedilum liemianum var. *primulinum*	**Paphiopedilum primulinum**
Paphiopedilum lindenii	**Phragmipedium lindenii**
Paphiopedilum lindleyanum	**Phragmipedium lindleyanum**
Paphiopedilum linii	**Paphiopedilum bullenianum**
Paphiopedilum longifolium	**Paphiopedilum longifolium**
Paphiopedilum lowii	
Paphiopedilum lowii f. *aureum*	**Paphiopedilum lowii**
Paphiopedilum lowii var. *aureum*	**Paphiopedilum lowii**
Paphiopedilum lowii var. *lynniae*	**Paphiopedilum lynniae**
Paphiopedilum lowii var. *richardianum*	**Paphiopedilum richardianum**
Paphiopedilum lynniae	
Paphiopedilum macfarlanei	**Paphiopedilium insigne**
Paphiopedilum malipoense	
Paphiopedilum malipoense f. *concolor*	**Paphiopedilum malipoense**
Paphiopedilum malipoense f. *tonnianum*	**Paphiopedilum malipoense**
Paphiopedilum malipoense f. *virescens*	**Paphiopedilum malipoense**
Paphiopedilum malipoense var. *hiepii*	**Paphiopedilum jackii var. hiepii**
Paphiopedilum malipoense var. *jackii*	**Paphiopedilum jackii**
Paphiopedilum markianum	**Paphiopedilum tigrinum**
Paphiopedilum mastersianum	
Paphiopedilum mastersianum var. mohrianum	
Paphiopedilum micranthum	

ALL NAMES	ACCEPTED NAMES
Paphiopedilum micranthum f. *alboflavum*	**Paphiopedilum micranthum**
Paphiopedilum micranthum f. *glanzeanum*	**Paphiopedilum micranthum**
Paphiopedilum micranthum subsp. *eburneum*	**Paphiopedilum micranthum**
Paphiopedilum micranthum var. *alboflavum*	**Paphiopedilum micranthum**
Paphiopedilum micranthum var. *glanzeanum*	**Paphiopedilum micranthum**
Paphiopedilum microchilum	**Paphiopedilum wardii**
Paphiopedilum mirabile	**Paphiopedilum vietnamense**
Paphiopedilum mohrianum	**Paphiopedilum mastersianum var. mohrianum**
Paphiopedilum moquetteanum	**Paphiopedilum glaucophyllum var. moquetteanum**
Paphiopedilum nicholsonianum	**Paphiopedilum rothschildianum**
Paphiopedilum nigritum	**Paphiopedilum barbatum**
Paphiopedilum niveum	
Paphiopedilum niveum f. *album*	**Paphiopedilum niveum**
Paphiopedilum ooii	
Paphiopedilum orbum	**Paphiopedilum callosum**
Paphiopedilum papuanum	
Paphiopedilum pardinum	**Paphiopedilum venustum**
Paphiopedilum parishii	
Paphiopedilum parishii var. *dianthum*	**Paphiopedilum parishii**
Paphiopedilum parnatanum	**Paphiopedilum usitanum**
Paphiopedilum paulistanum	**Phragmipedium vittatum**
Paphiopedilum pearcei	**Phragmipedium pearcei**
Paphiopedilum petri	**Paphiopedilum dayanum**
Paphiopedilum philippinense	
Paphiopedilum philippinense f. *alboflavum*	**Paphiopedilum philippinense**
Paphiopedilum philippinense f. *album*	**Paphiopedilum philippinense**
Paphiopedilum philippinense var. *cannartianum*	**Paphiopedilum philippinense**
Paphiopedilum philippinense var. roebelenii	
Paphiopedilum platyphyllum	
Paphiopedilum potentianum	**Paphiopedilum callosum var. potentianum**
Paphiopedilum praestans	**Paphiopedilum glanduliferum**
Paphiopedilum praestans subsp. *wilhelminae*	**Paphiopedilum wilhelminae**
Paphiopedilum praestans var. *wilhelminae*	**Paphiopedilum wilhelminae**
Paphiopedilum praestans var. *kimballianum*	**Paphiopedilum glanduliferum**
Paphiopedilum primulinum	
Paphiopedilum primulinum var. purpurascens	
Paphiopedilum puberulum	**Paphiopedilum appletonianum**
Paphiopedilum purpurascens	**Paphiopedilum javanicum var. virens**
Paphiopedilum purpuratum	
Paphiopedilum purpuratum var. *hainanense*	**Paphiopedilum purpuratum**
Paphiopedilum randsii	
Paphiopedilum reflexum	**Paphiopedilum callosum**
Paphiopedilum regnieri	**Paphiopedilum callosum**
Paphiopedilum reticulatum	**Phragmipedium reticulatum**
Paphiopedilum rhizomatosum	
Paphiopedilum richardianum	**Paphiopedilum lowii var. richardaianum**
Paphiopedilum robinsonii	**Paphiopedilum bullenianum**
Paphiopedilum robinsonii f. *viride*	**Paphiopedilum appletonianum**
Paphiopedilum roebelenii	**Paphiopedilum philippinense var. roebelenii**
Paphiopedilum roezlii	**Phragmipedium longifolium**
Paphiopedilum rothschildianum	
Paphiopedilum rothschildianum var. *elliottianum*	**Paphiopedilum rothschildianum**
Paphiopedilum saccopetalum	**Paphiopedilum hirsutissimum var. esquirolei**
Paphiopedilum sanderianum	
Paphiopedilum sangii	

ALL NAMES	ACCEPTED NAMES
Paphiopedilum sargentianum	**Phragmipedium lindleyanum**
Paphiopedilum schlimii	**Phragmipedium schlimii**
Paphiopedilum schoseri	
Paphiopedilum singchii	**Paphiopedilum hangianum**
Paphiopedilum sinicum	**Paphiopedilum purpuratum**
Paphiopedilum smaragdinum	**Paphiopedilum tigrinum**
Paphiopedilum socco	**Catasetum socco**
Paphiopedilum spicerianum	
Paphiopedilum spicerianum f. *immaculatum*	**Paphiopedilum spicerianum**
Paphiopedilum sriwanae	**Paphiopedilum argus**
Paphiopedilum stonei	
Paphiopedilum stonei subsp. *stictopetalum*	**Paphiopedilum spicerianum x stonei (hybrid)**
Paphiopedilum striatum	**Paphiopedilum wilhelminae**
Paphiopedilum sublaeve	**Paphiopedilum callosum var. warnerianum**
Paphiopedilum sugiyamanum	
Paphiopedilum sukhakulii	
Paphiopedilum sukhakulii f. *aureum*	**Paphiopedilum sukhakulii**
Paphiopedilum supardii	
Paphiopedilum superbiens	
Paphiopedilum superbiens f. *sanderae*	**Paphiopedilum superbiens**
Paphiopedilum superbiens subsp. *ciliolare*	**Paphiopedilum ciliolare**
Paphiopedilum super1biens var. *curtisii*	**Paphiopedilum superbiens**
Paphiopedilum superbiens var. *sanderae*	**Paphiopedilum superbiens**
Paphiopedilum thailandense	**Paphiopedilum callosum var. sublaeve**
Paphiopedilum tigrinum	
Paphiopedilum tonsum	
Paphiopedilum tonsum f. *alboviride*	**Paphiopedilum tonsum**
Paphiopedilum tonsum var. braemii	
Paphiopedilum topperi	**Paphiopedilum kolopakingii**
Paphiopedilum tortipetalum	**Paphiopedilum bullenianum**
Paphiopedilum tranlienianum	
Paphiopedilum tridentatum	**Paphiopedilum appletonianum**
Paphiopedilum urbanianum	
Paphiopedilum urbanianum f. *alboviride*	**Paphiopedilum urbanianum**
Paphiopedilum usitanum	
Paphiopedilum veitchianum	**Paphiopedilum superbiens**
Paphiopedilum vejvarutianum	**Paphiopedilum rhizomatosum**
Paphiopedilum venustum	
Paphiopedilum venustum f. *measuresianum*	**Paphiopedilum venustum**
Paphiopedilum venustum f. *pardinum*	**Paphiopedilum venustum**
Paphiopedilum venustum var. *bhutanensis*	**Paphiopedilum venustum**
Paphiopedilum venustum var. *pardinum*	**Paphiopedilum venustum**
Paphiopedilum venustum var. *rubrum*	**Paphiopedilum venustum**
Paphiopedilum venustum var. *teestaensis*	**Paphiopedilum venustum**
Paphiopedilum 'victoria'	**Paphiopedilum supardii**
Paphiopedilum victoria-mariae	
Paphiopedilum victoria-regina	
Paphiopedilum victoria-regina f. *purpurascens*	**Paphiopedilum primulinum var. purpurascens**
Paphiopedilum victoria-regina subsp. *chamberlainianum*	**Paphiopedilum victoria-regina**
Paphiopedilum victoria-regina subsp. *glaucophyllum*	**Paphiopedilum glaucophyllum**
Paphiopedilum victoria-regina subsp. *liemianum*	**Paphiopedilum liemianum**
Paphiopedilum victoria-regina subsp.var. *primulinum*	**Paphiopedilum primulinum**
Paphiopedilum victoria-regina var. *kalinae*	**Paphiopedilum victoria-regina**
Paphiopedilum victoria-regina var. *moquetteanum*	**Paphiopedilum glaucophyllum var. moquetteanum**
Paphiopedilum vietnamense	
Paphiopedilum villosum	

ALL NAMES	ACCEPTED NAMES
Paphiopedilum villosum f. *annamense*	**Paphiopedilum villosum var. annamense**
Paphiopedilum villosum f. *aureum*	**Paphiopedilum villosum**
Paphiopedilum villosum var. *affine*	**Paphiopedilum × affine (P. appletonianum × P. villosum)**
Paphiopedilum villosum var. annamense	
Paphiopedilum villosum var. boxallii	
Paphiopedilum villosum var. *gratrixianum*	**Paphiopedilum gratrixianum**
Paphiopedilum viniferum	**Paphiopedilum callosum**
Paphiopedilum violascens	
Paphiopedilum violascens var. *bougainvilleanum*	**Paphiopedilum bougainvilleanum**
Paphiopedilum violascens var. *gautierense*	**Paphiopedilum violascens**
Paphiopedilum violascens var. *saskianum*	**Paphiopedilum bougainvilleanum var. saskianum**
Paphiopedilum virens	**Paphiopedilum javanicum var. virens**
Paphiopedilum vittatum	**Phragmipedium vittatum**
Paphiopedilum volonteanum	**Paphiopedilum hookerae var. volonteanum**
Paphiopedilum volonteanum f. *sandowiae*	**Paphiopedilum hookerae var. volonteanum**
Paphiopedilum wallisii	**Phragmipedium wallisii**
Paphiopedilum wardii	
Paphiopedilum wardii f. *alboviride*	**Paphiopedilum wardii**
Paphiopedilum wardii var. *alboviride*	**Paphiopedilum wardii**
Paphiopedilum warscewiczianum	**Phragmipedium caudatum**
Paphiopedilum wenshanense	**Paphiopedilum bellatulum**
Paphiopedilum wentworthianum	
Paphiopedilum wilhelminae	
Paphiopedilum wolterianum	**Paphiopedilum appletonianum**
Paphiopedilum xerophyticum	**Mexipedium xerophyticum**
Paphiopedilum zieckianum	**Paphiopedilum papuanum**

PHRAGMIPEDIUM – LIST OF ALL NAMES / PHRAGMIPEDIUM – LISTE COMPLÈTE DES NOMS / PHRAGMIPEDIUM – LISTA DE TODOS LOS NOMBRES

ALL NAMES TOUS LES NOMS TODOS LOS NOMBRES	ACCEPTED NAME NOM ACCEPTÉS NOMBRES ACEPTADOS
Mexipedium xerophyticum	Formerly known as *Phragmipedium xerophyticum* this taxon remains on CITES Appendix I
Phragmipedium besseae	
Phragmipedium besseae f. *flavum*	**Phragmipedium besseae**
Phragmipedium besseae var. dalessandroi	
Phragmipedium besseae var. *flavum*	**Phragmipedium besseae**
Phragmipedium boissierianum	
Phragmipedium boissierianum var. czerwiakowianum	
Phragmipedium boissierianum var. reticulatum	
Phragmipedium brasilense	**Phragmipedium vittatum**
Phragmipedium cajamarcae	**Phragmipedium boissierianum**
Phragmipedium caricinum	
Phragmipedium caudatum	
Phragmipedium caudatum var. *lindenii*	**Phragmipedium lindenii subsp. lindenii**
Phragmipedium caudatum var. *wallisii*	**Phragmipedium lindenii subsp. wallisii**
Phragmipedium caudatum var. *warscewiczianum*	**Phragmipedium exstaminodium subsp. warscewiczii**
Phragmipedium chapadense	**Phragmipedium longifolium var. chapadense**
Phragmipedium christiansenianum	**Phragmipedium longifolium**
Phragmipedium czerwiakowianum	**Phragmipedium boissierianum var. czerwiakowianum**
Phragmipedium dalessandroi	**Phragmipedium besseae var. dalessandroi**
Phragmipedium dariense	**Phragmipedium longifolium**
Phragmipedium ecuadorense	**Phragmipedium pearcei**
Phragmipedium exstaminodium	
Phragmipedium exstaminodium subsp. exstaminodium	
Phragmipedium exstaminodium subsp. warscewiczii	
Phragmipedium fischeri	
Phragmipedium hartwegii	**Phragmipedium longifolium**
Phragmipedium hartwegii f. *baderi*	**Phragmipedium longifolium**
Phragmipedium hartwegii var. *baderi*	**Phragmipedium longifolium**
Phragmipedium hincksianum	**Phragmipedium longifolium**
Phragmipedium hirtzii	
Phragmipedium humboldtii	**Phragmipedium exstaminodium subsp. warscewiczii**
Phragmipedium humboldtii subsp. *exstaminodium*	**Phragmipedium exstaminodium subsp. exstaminodium**
Phragmipedium kaieteurum	**Phragmipedium lindleyanum var. kaiteurum**
Phragmipedium kovackii	
Phragmipedium klotzschianum	
Phragmipedium lindenii	
Phragmipedium lindenii subsp. lindenii	
Phragmipedium lindenii subsp. wallisii	
Phragmipedium lindleyanum	
Phragmipedium lindleyanum var. kaieteurum	
Phragmipedium longifolium	
Phragmipedium longifolium f. *gracile*	**Phragmipedium longifolium**
Phragmipedium longifolium f. *minutum*	**Phragmipedium longifolium**

ALL NAMES	ACCEPTED NAMES
Phragmipedium longifolium var. *darienense*	**Phragmipedium longifolium**
Phragmipedium longifolium var. *gracile*	**Phragmipedium longifolium**
Phragmipedium longifolium var. *hincksianum*	**Phragmipedium longifolium**
Phragmipedium longifolium var. *roezlii*	**Phragmipedium longifolium**
Phragmipedium pearcei	
Phragmipedium pearcei var. *ecuadorense*	**Phragmipedium pearcei**
Phragmipedium peruvianum	**Phragmipedium kovachii**
Phragmipedium portillae	**Phragmipedium richteri**
Phragmipedium popowii	**Phragmipedium exstaminodium subsp. warscewiczii**
Phragmipedium reticulatum	**Phragmipedium boissieranum**
Phragmipedium richteri	
Phragmipedium roezlii	**Phragmipedium longifolium**
Phragmipedium sargentianum	**Phragmipedium lindleyanum**
Phragmipedium schlimii	
Phragmipedium schlimii f. *albiflorum*	**Phragmipedium schlimii**
Phragmipedium schlimii var. *albiflorum*	**Phragmipedium schlimii**
Phragmipedium tetzlaffianum	
Phragmipedium vittatum	
Phragmipedium wallisii	**Phragmipedium lindenii subsp. wallisii**
Phragmipedium warszewiczianum	
Phragmipedium xerophyticum	**Mexipedium xerophyticum**